T0324914

Graduate Texts in Mathematics 42

Springer
New York
Berlin
Heidelberg
Barcelona
Hong Kong
London
Milan
Paris
Singapore
Tokyo

Graduate Texts in Mathematics

(continued after index)

Jean-Pierre Serre

Linear Representations of Finite Groups

Translated from the French by
Leonard L. Scott

Jean-Pierre Serre
Collège de France
75231 Paris Cedex 05
France

Leonard L. Scott
University of Virginia
Department of Mathematics
Charlottesville, Virginia 22903
USA

Mathematics Subject Classification: 20Cxx

Library of Congress Cataloging in Publication Data

Serre, Jean-Pierre.
Linear representations of finite groups.
(Graduate texts in mathematics ; 42)
Translation of Représentations linéaires des groupes finis, 2. ed.
Includes bibliographies and indexes.
1. Representations of groups. 2. Finite groups. I. Title. II. Series.
QA171.S5313 512'.2 76-12585

Translation of the French edition
Représentations linéaires des groupes finis, Paris: Hermann 1971

Printed on acid-free paper.

Printed and bound by Edwards Brothers, Inc., Ann Arbor, Michigan.
Printed in the United States of America.

9 8 7 6

ISBN 0-387-90190-6 SPIN 10834786
ISBN 3-540-90190-6

Springer-Verlag New York Berlin Heidelberg
A member of BertelsmannSpringer Science+Business Media GmbH

Preface

This book consists of three parts, rather different in level and purpose:

The first part was originally written for quantum chemists. It describes the correspondence, due to Frobenius, between linear representations and characters. This is a fundamental result, of constant use in mathematics as well as in quantum chemistry or physics. I have tried to give proofs as elementary as possible, using only the definition of a group and the rudiments of linear algebra. The examples (Chapter 5) have been chosen from those useful to chemists.

The second part is a course given in 1966 to second-year students of l'École Normale. It completes the first on the following points:

(a) degrees of representations and integrality properties of characters (Chapter 6);
(b) induced representations, theorems of Artin and Brauer, and applications (Chapters 7–11);
(c) rationality questions (Chapters 12 and 13).

The methods used are those of linear algebra (in a wider sense than in the first part): group algebras, modules, noncommutative tensor products, semisimple algebras.

The third part is an introduction to Brauer theory: passage from characteristic 0 to characteristic p (and conversely). I have freely used the language of abelian categories (projective modules, Grothendieck groups), which is well suited to this sort of question. The principal results are:

(a) The fact that the decomposition homomorphism is surjective: all irreducible representations in characteristic p can be lifted "virtually" (i.e., in a suitable Grothendieck group) to characteristic 0.
(b) The Fong–Swan theorem, which allows suppression of the word "virtually" in the preceding statement, provided that the group under consideration is p-solvable.

I have also given several applications to the Artin representations.

I take pleasure in thanking:

Gaston Berthier and Josiane Serre, who have authorized me to reproduce Part I, written for them and their students in *Quantum Chemistry*;

Yves Balasko, who drafted a first version of Part II from some lecture notes;

Alexandre Grothendieck, who has authorized me to reproduce Part III, which first appeared in his Séminaire de Géométrie Algébrique, I.H.E.S., 1965/66.

Contents

Part III
Introduction to Brauer Theory *113*

I

REPRESENTATIONS AND CHARACTERS

CHAPTER 1

Generalities on linear representations

1.1 Definitions

Let V be a vector space over the field **C** of complex numbers and let **GL**(V) be the group of *isomorphisms* of V onto itself. An element a of **GL**(V) is, by definition, a linear mapping of V into V which has an inverse a^{-1}; this inverse is linear. When V has a finite basis (e_i) of n elements, each linear map $a\colon V \to V$ is defined by a square matrix (a_{ij}) of order n. The coefficients a_{ij} are complex numbers; they are obtained by expressing the images $a(e_j)$ in terms of the basis (e_i):

$$a(e_j) = \sum_i a_{ij} e_i .$$

Saying that a is an isomorphism is equivalent to saying that the determinant $\det(a) = \det(a_{ij})$ of a is not zero. The group **GL**(V) is thus identifiable with the group of *invertible square matrices of order n.*

Suppose now G is a *finite* group, with identity element 1 and with composition $(s, t) \mapsto st$. A *linear representation* of G in V is a homomorphism ρ from the group G into the group **GL**(V). In other words, we associate with each element $s \in G$ an element $\rho(s)$ of **GL**(V) in such a way that we have the equality

$$\rho(st) = \rho(s) \cdot \rho(t) \quad \text{for } s, t \in G.$$

[We will also frequently write ρ_s instead of $\rho(s)$.] Observe that the preceding formula implies the following:

$$\rho(1) = 1, \quad \rho(s^{-1}) = \rho(s)^{-1}.$$

When ρ is given, we say that V is a *representation space* of G (or even simply, by abuse of language, a *representation* of G). In what follows, we restrict ourselves to the case where V has *finite dimension*. This is not a very severe restriction. Indeed, for most applications, one is interested in dealing with a *finite number of elements* x_i of V, and can always find a *subrepresentation* of V (in a sense defined later, cf. 1.3) of finite dimension, which contains the x_i: just take the vector subspace generated by the images $\rho_s(x_i)$ of the x_i.

Suppose now that V has finite dimension, and let n be its dimension; we say also that n is the degree of the representation under consideration. Let (e_i) be a basis of V, and let R_s be the matrix of ρ_s with respect to this basis. We have

$$\det(R_s) \neq 0, \qquad R_{st} = R_s \cdot R_t \quad \text{if } s, t \in G.$$

If we denote by $r_{ij}(s)$ the coefficients of the matrix R_s, the second formula becomes

$$r_{ik}(st) = \sum_j r_{ij}(s) \cdot r_{jk}(t).$$

Conversely, given invertible matrices $R_s = (r_{ij}(s))$ satisfying the preceding identities, there is a corresponding linear representation ρ of G in V; this is what it means to give a representation "in matrix form."

Let ρ and ρ' be two representations of the same group G in vector spaces V and V′. These representations are said to be *similar* (or *isomorphic*) if there exists a linear isomorphism τ: V → V′ which "transforms" ρ into ρ', that is, which satisfies the identity

$$\tau \circ \rho(s) = \rho'(s) \circ \tau \quad \text{for all } s \in G.$$

When ρ and ρ' are given in matrix form by R_s and R'_s respectively, this means that there exists an invertible matrix T such that

$$T \cdot R_s = R'_s \cdot T, \quad \text{for all } s \in G,$$

which is also written $R'_s = T \cdot R_s \cdot T^{-1}$. We can *identify* two such representations (by having each $x \in V$ correspond to the element $\tau(x) \in V'$); in particular, ρ and ρ' have the same degree.

1.2 Basic examples

(a) A representation *of degree* 1 of a group G is a homomorphism ρ: G → **C***, where **C*** denotes the multiplicative group of nonzero complex numbers. Since each element of G has finite order, the values $\rho(s)$ of ρ are roots of unity; in particular, we have $|\rho(s)| = 1$.

If we take $\rho(s) = 1$ for all $s \in G$, we obtain a representation of G which is called the *unit* (or *trivial*) representation.

(b) Let g be the order of G, and let V be a vector space of dimension g, with a basis $(e_t)_{t \in G}$ indexed by the elements t of G. For $s \in G$, let ρ_s be the linear map of V into V which sends e_t to e_{st}; this defines a linear representation, which is called the *regular representation* of G. Its degree is equal to the order of G. Note that $e_s = \rho_s(e_1)$; hence note that the images of e_1 form a basis of V. Conversely, let W be a representation of G containing a vector w such that the $\rho_s(w)$, $s \in G$, form a basis of W; then W is isomorphic to the regular representation (an isomorphism τ: V $\rightarrow$ W is defined by putting $\tau(e_s) = \rho_s(w)$).

(c) More generally, suppose that G acts on a finite set X. This means that, for each $s \in G$, there is given a permutation $x \mapsto sx$ of X, satisfying the identities

$$1x = x, \; s(tx) = (st)x \quad \text{if } s, t \in G, x \in X.$$

Let V be a vector space having a basis $(e_x)_{x \in X}$ indexed by the elements of X. For $s \in G$ let ρ_s be the linear map of V into V which sends e_x to e_{sx}; the linear representation of G thus obtained is called the *permutation representation* associated with X.

1.3 Subrepresentations

Let ρ: G $\rightarrow$ **GL**(V) be a linear representation and let W be a vector subspace of V. Suppose that W is *stable* under the action of G (we say also "invariant"), or in other words, suppose that $x \in W$ implies $\rho_s x \in W$ for all $s \in G$. The restriction ρ_s^W of ρ_s to W is then an isomorphism of W onto itself, and we have $\rho_{st}^W = \rho_s^W \cdot \rho_t^W$. Thus ρ^W: G $\rightarrow$ **GL**(W) is a linear representation of G in W; W is said to be a *subrepresentation* of V.

Example. Take for V the regular representation of G [cf. 1.2 (b)], and let W be the subspace of dimension 1 of V generated by the element $x = \sum_{s \in G} e_s$. We have $\rho_s x = x$ for all $s \in G$; consequently W is a subrepresentation of V, isomorphic to the unit representation. (We will determine in 2.4 all the subrepresentations of the regular representation.)

Before going further, we recall some concepts from linear algebra. Let V be a vector space, and let W and W′ be two subspaces of V. The space V is said to be the *direct sum* of W and W′ if each $x \in V$ can be written uniquely in the form $x = w + w'$, with $w \in W$ and $w' \in W'$; this amounts to saying that the intersection $W \cap W'$ of W and W′ is 0 and that $\dim(V) = \dim(W) + \dim(W')$. We then write $V = W \oplus W'$ and say that W′ is a complement of W in V. The mapping p which sends each $x \in V$ to its component $w \in W$ is called the *projection* of V onto W associated with the decomposition $V = W \oplus W'$; the image of p is W, and $p(x) = x$ for $x \in W$; conversely if p is a linear map of V into itself satisfying these two properties, one checks that V is the direct sum of W and the *kernel* W′ of p

(the set of x such that $px = 0$). A bijective correspondence is thus established between the *projections* of V onto W and the *complements* of W in V.

We return now to subrepresentations:

Theorem 1. *Let* ρ: G $\rightarrow$ **GL**(V) *be a linear representation of* G *in* V *and let* W *be a vector subspace of* V *stable under* G. *Then there exists a complement* W^0 *of* W *in* V *which is stable under* G.

Let W′ be an arbitrary complement of W in V, and let p be the corresponding projection of V onto W. Form the average p^0 of the conjugates of p by the elements of G:

$$p^0 = \frac{1}{g} \sum_{t \in G} \rho_t \cdot p \cdot \rho_t^{-1} \qquad (g \text{ being the order of G}).$$

Since p maps V into W and ρ_t preserves W we see that p^0 maps V into W; we have $\rho_t^{-1} x \in W$ for $x \in W$, whence

$$p \cdot \rho_t^{-1} x = \rho_t^{-1} x, \qquad \rho_t \cdot p \cdot \rho_t^{-1} x = x, \qquad \text{and} \qquad p^0 x = x.$$

Thus p^0 is a projection of V onto W, corresponding to some complement W^0 of W. We have moreover

$$\rho_s \cdot p^0 = p^0 \cdot \rho_s \quad \text{for all } s \in G.$$

Indeed, computing $\rho_s \cdot p^0 \cdot \rho_s^{-1}$, we find:

$$\rho_s \cdot p^0 \cdot \rho_s^{-1} = \frac{1}{g} \sum_{t \in G} \rho_s \cdot \rho_t \cdot p \cdot \rho_t^{-1} \cdot \rho_s^{-1} = \frac{1}{g} \sum_{t \in G} \rho_{st} \cdot p \cdot \rho_{st}^{-1} = p^0.$$

If now $x \in W^0$ and $s \in G$ we have $p^0 x = 0$, hence $p^0 \cdot \rho_s x = \rho_s \cdot p^0 x = 0$, that is, $\rho_s x \in W^0$, which shows that W^0 is stable under G, and completes the proof. □

Remark. Suppose that V is endowed with a *scalar product* $(x|y)$ satisfying the usual conditions: linearity in x, semilinearity in y, and $(x|x) > 0$ if $x \neq 0$. Suppose that this scalar product is *invariant* under G, i.e., that $(\rho_s x|\rho_s y) = (x|y)$; we can always reduce to this case by replacing $(x|y)$ by $\sum_{t \in G} (\rho_t x|\rho_t y)$. Under these hypotheses the *orthogonal complement* W^0 *of* W in V is a complement of W stable under G; another proof of theorem 1 is thus obtained. Note that the invariance of the scalar product $(x|y)$ means that, if (e_i) is an orthonormal basis of V, the matrix of ρ_s with respect to this basis is a *unitary matrix*.

Keeping the hypothesis and notation of theorem 1, let $x \in V$ and let w and w^0 be its projections on W and W^0. We have $x = w + w^0$, whence $\rho_s x = \rho_s w + \rho_s w^0$, and since W and W^0 are stable under G, we have $\rho_s w \in W$ and $\rho_s w^0 \in W^0$; thus $\rho_s w$ and $\rho_s w^0$ are the projections of $\rho_s x$. It follows the representations W and W^0 determine the representation V.

We say that V is the direct sum of W and W^0, and write $V = W \oplus W^0$. An element of V is identified with a pair (w, w^0) with $w \in W$ and $w^0 \in W^0$. If W and W^0 are given in matrix form by R_s and R_s^0, $W \oplus W^0$ is given in matrix form by

$$\begin{pmatrix} R_s & 0 \\ 0 & R_s^0 \end{pmatrix}.$$

The direct sum of an arbitrary finite number of representations is defined similarly.

1.4 Irreducible representations

Let $\rho\colon G \to \mathbf{GL}(V)$ be a linear representation of G. We say that it is *irreducible* or *simple* if V is not 0 and if no vector subspace of V is stable under G, except of course 0 and V. By theorem 1, this second condition is equivalent to saying V *is not the direct sum of two representations* (except for the trivial decomposition $V = 0 \oplus V$). A representation of degree 1 is evidently irreducible. We will see later (3.1) that each nonabelian group possesses at least one irreducible representation of degree $\geqslant 2$.

The irreducible representations are used to construct the others by means of the direct sum:

Theorem 2. *Every representation is a direct sum of irreducible representations.*

Let V be a linear representation of G. We proceed by induction on $\dim(V)$. If $\dim(V) = 0$, the theorem is obvious (0 is the direct sum of the *empty family* of irreducible representations). Suppose then $\dim(V) \geqslant 1$. If V is irreducible, there is nothing to prove. Otherwise, because of th. 1, V can be decomposed into a direct sum $V' \oplus V''$ with $\dim(V') < \dim(V)$ and $\dim(V'') < \dim(V)$. By the induction hypothesis V' and V'' are direct sums of irreducible representations, and so the same is true of V. □

Remark. Let V be a representation, and let $V = W_1 \oplus \cdots \oplus W_k$ be a decomposition of V into a direct sum of irreducible representations. We can ask if this decomposition is *unique*. The case where all the ρ_s are equal to 1 shows that this is not true in general (in this case the W_i are lines, and we have a plethora of decompositions of a vector space into a direct sum of lines). Nevertheless, we will see in 2.3 that the *number* of W_i isomorphic to a given irreducible representation does not depend on the chosen decomposition.

1.5 Tensor product of two representations

Along with the direct sum operation (which has the formal properties of an addition), there is a "multiplication": the *tensor product*, sometimes called the *Kronecker* product. It is defined as follows:

To begin with, let V_1 and V_2 be two vector spaces. A space W furnished with a map $(x_1, x_2) \mapsto x_1 \cdot x_2$ of $V_1 \times V_2$ into W, is called the tensor product of V_1 and V_2 if the two following conditions are satisfied:

(i) $x_1 \cdot x_2$ is linear in each of the variables x_1 and x_2.
(ii) If (e_{i_1}) is a basis of V_1 and (e_{i_2}) is a basis of V_2, the family of products $e_{i_1} \cdot e_{i_2}$ is a basis of W.

It is easily shown that such a space exists, and is unique (up to isomorphism); it is denoted $V_1 \otimes V_2$. Condition (ii) shows that

$$\dim(V_1 \otimes V_2) = \dim(V_1) \cdot \dim(V_2).$$

Now let $\rho^1: G \to \mathbf{GL}(V_1)$ and $\rho^2: G \to \mathbf{GL}(V_2)$ be two linear representations of a group G. For $s \in G$, define an element ρ_s of $\mathbf{GL}(V_1 \otimes V_2)$ by the condition:

$$\rho_s(x_1 \cdot x_2) = \rho_s^1(x_1) \cdot \rho_s^2(x_2) \quad \text{for } x_1 \in V_1, x_2 \in V_2.$$

[The existence and uniqueness of ρ_s follows easily from conditions (i) and (ii).] We write:

$$\rho_s = \rho_s^1 \otimes \rho_s^2.$$

The ρ_s define a linear representation of G in $V_1 \otimes V_2$ which is called the *tensor product* of the given representations.

The matrix translation of this definition is the following: let (e_{i_1}) be a basis for V_1, let $r_{i_1 j_1}(s)$ be the matrix of ρ_s^1 with respect to this basis, and define (e_{i_2}) and $r_{i_2 j_2}(s)$ in the same way. The formulas:

$$\rho_s^1(e_{j_1}) = \sum_{i_1} r_{i_1 j_1}(s) \cdot e_{i_1}, \qquad \rho_s^2(e_{j_2}) = \sum_{i_2} r_{i_2 j_2}(s) \cdot e_{i_2}$$

imply:

$$\rho_s(e_{j_1} \cdot e_{j_2}) = \sum_{i_1, i_2} r_{i_1 j_1}(s) \cdot r_{i_2 j_2}(s) \cdot e_{i_1} \cdot e_{i_2}.$$

Accordingly the matrix of ρ_s is $(r_{i_1 j_1}(s) \cdot r_{i_2 j_2}(s))$; it is the *tensor product* of the matrices of ρ_s^1 and ρ_s^2.

The tensor product of two irreducible representations is not in general irreducible. It decomposes into a direct sum of irreducible representations which can be determined by means of character theory (cf. 2.3).

In quantum chemistry, the tensor product often appears in the following way: V_1 and V_2 are two spaces of functions stable under G, with respective bases (ϕ_{i_1}) and (ψ_{i_2}), and $V_1 \otimes V_2$ is the vector space generated by the products $\phi_{i_1} \cdot \psi_{i_2}$, these products being linearly independent. This last condition is essential. Here are two particular cases where it is satisfied:

(1) The ϕ's depend only on certain variables $(x, x', \ldots)$ and the ψ's on variables $(y, y', \ldots)$ independent from the first.
(2) The space V_1 (or V_2) has a basis consisting of a single function ϕ, this function does not vanish identically in any region; the space V_1 is then of dimension 1.

1.6 Symmetric square and alternating square

Suppose that the representations V_1 and V_2 are identical to the same representation V, so that $V_1 \otimes V_2 = V \otimes V$. If (e_i) is a basis of V, let θ be the automorphism of $V \otimes V$ such that

$$\theta(e_i \cdot e_j) = e_j \cdot e_i \quad \text{for all pairs } (i,j).$$

It follows from this that $\theta(x \cdot y) = y \cdot x$ for $x, y \in V$, hence that θ is independent of the chosen basis (e_i); moreover $\theta^2 = 1$. The space $V \otimes V$ then decomposes into a direct sum

$$V \otimes V = \mathbf{Sym}^2(V) \oplus \mathbf{Alt}^2(V),$$

where $\mathbf{Sym}^2(V)$ is the set of elements $z \in V \otimes V$ such that $\theta(z) = z$ and $\mathbf{Alt}^2(V)$ is the set of elements $z \in V \otimes V$ such that $\theta(z) = -z$. The elements $(e_i \cdot e_j + e_j \cdot e_i)_{i \leqslant j}$ form a basis of $\mathbf{Sym}^2(V)$, and the elements $(e_i \cdot e_j - e_j \cdot e_i)_{i<j}$ form a basis of $\mathbf{Alt}^2(V)$. We have

$$\dim \mathbf{Sym}^2(V) = \frac{n(n+1)}{2}, \qquad \dim \mathbf{Alt}^2(V) = \frac{n(n-1)}{2}$$

if $\dim V = n$.

The subspaces $\mathbf{Sym}^2(V)$ and $\mathbf{Alt}^2(V)$ are stable under G, and thus define representations called respectively the *symmetric square* and *alternating square* of the given representation.

CHAPTER 2

Character theory

2.1 The character of a representation

Let V be a vector space having a basis (e_i) of n elements, and let a be a linear map of V into itself, with matrix (a_{ij}). By the *trace* of a we mean the scalar

$$\mathrm{Tr}(a) = \sum_i a_{ii}.$$

It is the *sum of the eigenvalues of a* (counted with their multiplicities), and does not depend on the choice of basis (e_i).

Now let $\rho\colon \mathrm{G} \to \mathbf{GL}(\mathrm{V})$ be a linear representation of a finite group G in the vector space V. For each $s \in \mathrm{G}$, put:

$$\chi_\rho(s) = \mathrm{Tr}(\rho_s).$$

The complex valued function χ_ρ on G thus obtained is called the *character* of the representation ρ; the importance of this function comes primarily from the fact that it *characterizes* the representation ρ (cf. 2.3).

Proposition 1. *If χ is the character of a representation ρ of degree n, we have*:

(i) $\chi(1) = n$,
(ii) $\chi(s^{-1}) = \chi(s)^*$ *for* $s \in \mathrm{G}$,
(iii) $\chi(tst^{-1}) = \chi(s)$ *for* $s, t \in \mathrm{G}$.

(If $z = x + iy$ is a complex number, we denote the conjugate $x - iy$ either by z^* or $\bar{z}$.)

We have $\rho(1) = 1$, and $\mathrm{Tr}(1) = n$ since V has dimension n; hence (i).

For (ii) we observe that ρ_s has finite order; consequently the same is true of its eigenvalues $\lambda_1, \ldots, \lambda_n$ and so these have absolute value equal to 1 (this is also a consequence of the fact that ρ_s can be defined by a unitary matrix, cf. 1.3). Thus

$$\chi(s)^* = \mathrm{Tr}(\rho_s)^* = \sum \lambda_i^* = \sum \lambda_i^{-1} = \mathrm{Tr}(\rho_s^{-1}) = \mathrm{Tr}(\rho_{s^{-1}}) = \chi(s^{-1}).$$

Formula (iii) can also be written $\chi(vu) = \chi(uv)$, putting $u = ts$, $v = t^{-1}$; hence it follows from the well known formula

$$\mathrm{Tr}(ab) = \mathrm{Tr}(ba),$$

valid for two arbitrary linear mappings a and b of V into itself. □

Remark. A function f on G satisfying identity (iii), or what amounts to the same thing, $f(uv) = f(vu)$, is called a *class function* We will see in 2.5 that each class function is a linear combination of characters.

Proposition 2. *Let $\rho^1 \colon G \to \mathbf{GL}(V_1)$ and $\rho^2 \colon G \to \mathbf{GL}(V_2)$ be two linear representations of* G, *and let χ_1 and χ_2 be their characters. Then*:

(i) *The character χ of the direct sum representation* $V_1 \oplus V_2$ *is equal to* $\chi_1 + \chi_2$.
(ii) *The character ψ of the tensor product representation* $V_1 \otimes V_2$ *is equal to* $\chi_1 \cdot \chi_2$.

Let us be given ρ^1 and ρ^2 in matrix form: R_s^1, R_s^2. The representation $V_1 \oplus V_2$ is then given by

$$R_s = \begin{pmatrix} R_s^1 & 0 \\ 0 & R_s^2 \end{pmatrix}$$

whence $\mathrm{Tr}(R_s) = \mathrm{Tr}(R_s^1) + \mathrm{Tr}(R_s^2)$, that is $\chi(s) = \chi_1(s) + \chi_2(s)$.
We proceed likewise for (ii): with the notation of 1.5, we have

$$\chi_1(s) = \sum_{i_1} r_{i_1 i_1}(s), \qquad \chi_2(s) = \sum_{i_2} r_{i_2 i_2}(s),$$

$$\psi(s) = \sum_{i_1, i_2} r_{i_1 i_1}(s) r_{i_2 i_2}(s) = \chi_1(s) \cdot \chi_2(s). \qquad \square$$

Proposition 3. *Let $\rho \colon G \to \mathbf{GL}(V)$ be a linear representation of* G, *and let χ be its character. Let χ_σ^2 be the character of the symmetric square* $\mathbf{Sym}^2(V)$ *of* V (cf. 1.6), *and let χ_α^2 be that of* $\mathbf{Alt}^2(V)$. *For each $s \in G$, we have*

$$\chi_\sigma^2(s) = \frac{1}{2}(\chi(s)^2 + \chi(s^2))$$

$$\chi_\alpha^2(s) = \frac{1}{2}(\chi(s)^2 - \chi(s^2))$$

and $\chi_\sigma^2 + \chi_\alpha^2 = \chi^2$.

Let $s \in G$. A basis (e_i) of V can be chosen consisting of *eigenvectors* for ρ_s; this follows for example from the fact that ρ_s can be represented by a *unitary* matrix, cf. 1.3. We have then $\rho_s e_i = \lambda_i e_i$ with $\lambda_i \in \mathbf{C}$, and so

$$\chi(s) = \sum \lambda_i, \qquad \chi(s^2) = \sum \lambda_i^2.$$

On the other hand, we have

$$(\rho_s \otimes \rho_s)(e_i \cdot e_j + e_j \cdot e_i) = \lambda_i \lambda_j \cdot (e_i \cdot e_j + e_j \cdot e_i),$$

$$(\rho_s \otimes \rho_s)(e_i \cdot e_j - e_j \cdot e_i) = \lambda_i \lambda_j \cdot (e_i \cdot e_j - e_j \cdot e_i),$$

hence

$$\chi_\sigma^2(s) = \sum_{i \leqslant j} \lambda_i \lambda_j = \sum \lambda_i^2 + \sum_{i<j} \lambda_i \lambda_j = \frac{1}{2}(\sum \lambda_i)^2 + \frac{1}{2} \sum \lambda_i^2$$

$$\chi_\alpha^2(s) - \sum_{i<j} \lambda_i \lambda_j = \frac{1}{2}(\sum \lambda_i)^2 - \frac{1}{2} \sum \lambda_i^2.$$

The proposition follows.

(Observe the equality $\chi_\sigma^2 + \chi_\alpha^2 = \chi^2$, which reflects the fact that $V \otimes V$ is the *direct sum* of $\mathbf{Sym}^2(V)$ and $\mathbf{Alt}^2(V)$). □

Exercises

2.1. Let χ and χ' be the characters of two representations. Prove the formulas:

$$(\chi + \chi')_\sigma^2 = \chi_\sigma^2 + \chi_\sigma'^2 + \chi\chi',$$

$$(\chi + \chi')_\alpha^2 = \chi_\alpha^2 + \chi_\alpha'^2 + \chi\chi'.$$

2.2. Let X be a finite set on which G acts, let ρ be the corresponding permutation representation [cf. 1.2, example (c)], and χ_X be the character of ρ. Let $s \in G$; show that $\chi_X(s)$ is the number of elements of X fixed by s.

2.3. Let $\rho: G \to \mathbf{GL}(V)$ be a linear representation with character χ and let V' be the dual of V, i.e., the space of linear forms on V. For $x \in V$, $x' \in V'$ let $\langle x, x' \rangle$ denote the value of the linear form x' at x. Show that there exists a unique linear representation $\rho': G \to \mathbf{GL}(V')$, such that

$$\langle \rho_s x, \rho'_s x' \rangle = \langle x, x' \rangle \quad \text{for } s \in G,\ x \in V,\ x' \in V'.$$

This is called the *contragredient* (or *dual*) representation of ρ; its character is χ^*.

2.4. Let $\rho_1: G \to \mathbf{GL}(V_1)$ and $\rho_2: G \to \mathbf{GL}(V_2)$ be two linear representations with characters χ_1 and χ_2. Let $W = \mathrm{Hom}(V_1, V_2)$, the vector space of linear mappings $f: V_1 \to V_2$. For $s \in G$ and $f \in W$ let $\rho_s f = \rho_{2,s} \circ f \circ \rho_{1,s}^{-1}$; so $\rho_s f \in W$. Show that this defines a linear representation $\rho: G \to \mathbf{GL}(W)$, and that its character is $\chi_1^* \cdot \chi_2$. This representation is isomorphic to $\rho'_1 \otimes \rho_2$, where ρ_1' is the contragredient of ρ_1, cf. ex. 2.3.

2.2 Schur's lemma; basic applications

Proposition 4(*Schur's lemma*).*Let* $\rho^1: \mathrm{G} \to \mathbf{GL}(\mathrm{V}_1)$ *and* $\rho^2: \mathrm{G} \to \mathbf{GL}(\mathrm{V}_2)$ *be two irreducible representations of* G, *and let f be a linear mapping of* V_1 *into* V_2 *such that* $\rho_s^2 \circ f = f \circ \rho_s^1$ *for all* $s \in \mathrm{G}$. *Then*:

(1) *If* ρ^1 *and* ρ^2 *are not isomorphic, we have* $f = 0$.

(2) *If* $\mathrm{V}_1 = \mathrm{V}_2$ *and* $\rho^1 = \rho^2$, *f is a homothety* (*i.e., a scalar multiple of the identity*).

The case $f = 0$ is trivial. Suppose now $f \neq 0$ and let W_1 be *its kernel* (that is, the set of $x \in \mathrm{V}_1$ such that $fx = 0$). For $x \in \mathrm{W}_1$ we have $f\rho_s^1 x = \rho_s^2 fx = 0$, whence $\rho_s^1 \chi \in \mathrm{W}_1$, and W_1 is stable under G. Since V_1 is irreducible, W_1 is equal to V_1 or 0; the first case is excluded, as it implies $f = 0$. The same argument shows that the *image* W_2 of f (the set of fx, for $x \in \mathrm{V}_1$) is equal to V_2. The two properties $\mathrm{W}_1 = 0$ and $\mathrm{W}_2 = \mathrm{V}_2$ show that f is an isomorphism of V_1 onto V_2, which proves assertion (1).

Suppose now that $\mathrm{V}_1 = \mathrm{V}_2$, $\rho^1 = \rho^2$, and let λ be an *eigenvalue* of f: there exists at least one, since the field of scalars is the field of complex numbers. Put $f' = f - \lambda$. Since λ is an eigenvalue of f, the kernel of f' is $\neq 0$; on the other hand, we have $\rho_s^2 \circ f' = f' \circ \rho_s^1$. The first part of the proof shows that these properties are possible only if $f' = 0$, that is, if f is equal to λ. □

Let us keep the hypothesis that V_1 and V_2 are irreducible, and denote by g the *order* of the group G.

Corollary 1. *Let h be a linear mapping of* V_1 *into* V_2, *and put*:

$$h^0 = \frac{1}{g} \sum_{t \in \mathrm{G}} (\rho_t^2)^{-1} h \rho_t^1 .$$

Then:

(1) *If* ρ^1 *and* ρ^2 *are not isomorphic, we have* $h^0 = 0$.

(2) *If* $\mathrm{V}_1 = \mathrm{V}_2$ *and* $\rho^1 = \rho^2$, h^0 *is a homothety of ratio* $(1/n)\mathrm{Tr}(h)$, *with* $n = \dim(\mathrm{V}_1)$.

We have $\rho_s^2 h^0 = h^0 \rho_s^1$. Indeed:

$$(\rho_s^2)^{-1} h^0 \rho_s^1 = \frac{1}{g} \sum_{t \in \mathrm{G}} (\rho_s^2)^{-1} (\rho_t^2)^{-1} h \rho_t^1 \rho_s^1$$

$$= \frac{1}{g} \sum_{t \in \mathrm{G}} (\rho_{ts}^2)^{-1} h \rho_{ts}^1 = h^0 .$$

Applying prop. 4 to $f = h^0$, we see in case (1) that $h^0 = 0$, and in case (2) that h^0 is equal to a scalar λ. Moreover, in the latter case, we have:

$$\mathrm{Tr}(h^0) = \frac{1}{g} \sum_{t \in \mathrm{G}} \mathrm{Tr}((\rho_t^1)^{-1} h \rho_t^1) = \mathrm{Tr}(h),$$

and since $\mathrm{Tr}(\lambda) = n \cdot \lambda$, we get $\lambda = (1/n)\mathrm{Tr}(h)$. □

Now we rewrite corollary 1 assuming that ρ^1 and ρ^2 are given *in matrix form*:

$$\rho_t^1 = (r_{i_1 j_1}(t)),\ \rho_t^2 = (r_{i_2 j_2}(t)).$$

The linear mapping h is defined by a matrix $(x_{i_2 i_1})$ and likewise h^0 is defined by $(x^0_{i_2 i_1})$. We have by definition of h^0:

$$x^0_{i_2 i_1} = \frac{1}{g} \sum_{t, j_1, j_2} r_{i_2 j_2}(t^{-1}) x_{j_2 j_1} r_{j_1 i_1}(t).$$

The right hand side is a linear form with respect to $x_{j_2 j_1}$; in case (1) this form vanishes for all systems of values of the $x_{j_2 j_1}$; thus its coefficients are zero. Whence:

Corollary 2. *In case* (1), *we have*:

$$\frac{1}{g} \sum_{t \in \mathrm{G}} r_{i_2 j_2}(t^{-1}) r_{j_1 i_1}(t) = 0$$

for arbitrary i_1, i_2, j_1, j_2.

In case (2) we have similarly $h^0 = \lambda$, i.e., $x^0_{i_2 i_1} = \lambda \delta_{i_2 i_1}$ ($\delta_{i_2 i_1}$ denotes the Kronecker symbol, equal to 1 if $i_1 = i_2$ and 0 otherwise), with $\lambda = (1/n)\mathrm{Tr}(h)$, that is, $\lambda = (1/n) \sum \delta_{j_2 j_1} x_{j_2 j_1}$. Hence the equality:

$$\frac{1}{g} \sum_{t, j_1, j_2} r_{i_2 j_2}(t^{-1}) x_{j_2 j_1} r_{j_1 i_1}(t) = \frac{1}{n} \sum_{j_1, j_2} \delta_{i_2 i_1} \delta_{j_2 j_1} x_{j_2 j_1}.$$

Equating coefficients of the $x_{j_2 j_1}$, we obtain as above:

Corollary 3. *In case* (2) *we have*:

$$\frac{1}{g} \sum_{t \in \mathrm{G}} r_{i_2 j_2}(t^{-1}) r_{j_1 i_1}(t) = \frac{1}{n} \delta_{i_2 i_1} \delta_{j_2 j_1} = \begin{cases} 1/n & \textit{if } i_1 = i_2 \textit{ and } j_1 = j_2 \\ 0 & \textit{otherwise}. \end{cases}$$

Remarks

(1) If ϕ and ψ are functions on G, set

$$\langle \phi, \psi \rangle = \frac{1}{g} \sum_{t \in \mathrm{G}} \phi(t^{-1}) \psi(t) = \frac{1}{g} \sum_{t \in \mathrm{G}} \phi(t) \psi(t^{-1}).$$

We have $\langle \phi, \psi \rangle = \langle \psi, \phi \rangle$. Moreover $\langle \phi, \psi \rangle$ is linear in ϕ and in ψ. With this notation, corollaries 2 and 3 become, respectively

$$\langle r_{i_2 j_2}, r_{j_1 i_1} \rangle = 0 \quad \text{and} \quad \langle r_{i_2 j_2}, r_{j_1 i_1} \rangle = \frac{1}{n} \delta_{i_2 i_1} \delta_{j_2 j_1}.$$

(2) Suppose that the matrices $(r_{ij}(t))$ are *unitary* (this can be realized by a suitable choice of basis, cf. 1.3). We have then $r_{ij}(t^{-1}) = r_{ji}(t)^*$ and corollaries 2 and 3 are just *orthogonality relations* for the scalar product $(\phi|\psi)$ defined in the following section.

2.3 Orthogonality relations for characters

We begin with a notation. If ϕ and ψ are two complex-valued functions on G, put

$$(\phi|\psi) = \frac{1}{g} \sum_{t \in G} \phi(t)\psi(t)^*, \quad g \text{ being the order of } G .$$

This is a *scalar product*: it is linear in ϕ, semilinear in ψ, and we have $(\phi|\phi) > 0$ for all $\phi \neq 0$.

If $\check{\psi}$ is the function defined by the formula $\check{\psi}(t) = \psi(t^{-1})^*$, we have

$$(\phi|\psi) = \frac{1}{g} \sum_{t \in G} \phi(t)\check{\psi}(t^{-1}) = \langle \phi, \check{\psi} \rangle,$$

cf. 2.2, remark 1. In particular, if χ is the *character* of a representation of G, we have $\check{\chi} = \chi$ (prop. 1), so that $(\phi|\chi) = \langle \phi, \chi \rangle$ for all functions ϕ on G. So we can use at will $(\phi|\chi)$ or $\langle \phi, \chi \rangle$, so long as we are concerned with characters.

Theorem 3

(i) *If χ is the character of an irreducible representation, we have* $(\chi|\chi) = 1$ (i.e., χ is "of norm 1").

ii) *If χ and χ' are the characters of two nonisomorphic irreducible representations, we have* $(\chi|\chi') = 0$ (i.e. χ and χ' are orthogonal).

Let ρ be an irreducible representation with character χ, given in matrix form $\rho_t = (r_{ij}(t))$. We have $\chi(t) = \sum r_{ii}(t)$, hence

$$(\chi|\chi) = \langle \chi, \chi \rangle = \sum_{i,j} \langle r_{ii}, r_{jj} \rangle.$$

But according to cor. 3 to prop. 4, we have $\langle r_{ii}, r_{jj} \rangle = \delta_{ij}/n$, where n is the degree of ρ. Thus

$$(\chi|\chi) = \Big(\sum_{i,j} \delta_{ij}\Big)/n = n/n = 1,$$

since the indices i,j each take n values. (ii) is proved in the same way, by applying cor. 2 instead of cor. 3. □

Remark. A character of an irreducible representation is called an *irreducible character.* Theorem 3 shows that the irreducible characters form an orthonormal system; this result will be completed later (2.5, th. 6).

Theorem 4. *Let* V *be a linear representation of* G, *with character* ϕ, *and suppose* V *decomposes into a direct sum of irreducible representations*:

$$V = W_1 \oplus \cdots \oplus W_k.$$

Then, if W *is an irreducible representation with character* χ, *the number of* W_i *isomorphic to* W *is equal to the scalar product* $(\phi|\chi) = \langle\phi, \chi\rangle$.

Let χ_i be the character of W_i. By prop. 2, we have

$$\phi = \chi_1 + \cdots + \chi_k.$$

Thus $(\phi|\chi) = (\chi_1|\chi) + \cdots + (\chi_k|\chi)$. But, according to the preceeding theorem, $(\chi_i|\chi)$ is equal to 1 or 0, depending on whether W_i is, or is not, isomorphic to W. The result follows. □

Corollary 1. *The number of* W_i *isomorphic to* W *does not depend on the chosen decomposition.*

(This number is called the "number of times that W occurs in V", or the "number of times that W is contained in V.")

Indeed, $(\phi|\chi)$ does not depend on the decomposition. □

Remark. It is in this sense that one can say that there is uniqueness in the decomposition of a representation into irreducible representations. We shall return to this in 2.6.

Corollary 2. *Two representations with the same character are isomorphic.*

Indeed, cor. 1 shows that they contain each given irreducible representation the same number of times.

The above results reduce the study of representations to that of their characters. If $\chi_1, \ldots, \chi_h$ are the distinct irreducible characters of G, and if $W_1, \ldots, W_h$ denote corresponding representations, each representation V is isomorphic to a direct sum

$$V = m_1 W_1 \oplus \cdots \oplus m_h W_h \quad m_i \text{ integers } \geqslant 0.$$

The character ϕ of V is equal to $m_1\chi_1 + \cdots + m_h\chi_h$, and we have $m_i = (\phi|\chi_i)$. [This applies notably to the tensor product $W_i \otimes W_j$ of two irreducible representations, and shows that the product $\chi_i \cdot \chi_j$ decomposes into $\chi_i\chi_j = \sum m_{ij}^k \chi_k$, the m_{ij}^k being integers $\geqslant 0$.] The orthogonality relations among the χ_i imply in addition:

$$(\phi|\phi) = \sum_{i=1}^{i=h} m_i^2,$$

whence:

Theorem 5. *If ϕ is the character of a representation* V, $(\phi|\phi)$ *is a positive integer and we have* $(\phi|\phi) = 1$ *if and only if* V *is irreducible.*

Indeed, $\sum m_i^2$ is only equal to 1 if one of the m_i's is equal to 1 and the others to 0, that is, if V is isomorphic to one of the W_i. □

We obtain thus a very convenient *irreducibility criterion.*

EXERCISES

2.5. Let ρ be a linear representation with character χ. Show that the number of times that ρ contains the unit representation is equal to $(\chi|1) = (1/g) \sum_{s \in G} \chi(s)$.

2.6. Let X be a finite set on which G acts, let ρ be the corresponding permutation representation (1.2) and let χ be its character.

(a) The set Gx of images under G of an element $x \in$ X is called an *orbit.* Let c be the number of distinct orbits. Show that c is equal to the number of times that ρ contains the unit representation 1; deduce from this that $(\chi|1) = c$. In particular, if G is transitive (i.e., if $c = 1$), ρ can be decomposed into $1 \oplus \theta$ and θ does not contain the unit representation. If ψ is the character of θ, we have $\chi = 1 + \psi$ and $(\psi|1) = 0$.

(b) Let G act on the product X × X of X by itself by means of the formula $s(x,y) = (sx, sy)$. Show that the character of the corresponding permutation representation is equal to χ^2.

(c) Suppose that G is *transitive* on X and that X has at least two elements. We say that G is doubly transitive if, for all $x, y, x', y' \in$ X with $x \neq y$ and $x' \neq y'$, there exists $s \in$ G such that $x' = sx$ and $y' = sy$. Prove the equivalence of the following properties:

(i) G is doubly transitive.

(ii) The action of G on X × X has two orbits, the diagonal and its complement.

(iii) $(\chi^2|1) = 2$.

(iv) The representation θ defined in (a) is irreducible.

[The equivalence (i) ⇔ (ii) is immediate; (ii) ⇔ (iii) follows from (a) and (b). If ψ is the character of θ, we have $1 + \psi = \chi$ and $(1|1) = 1$, $(\psi|1) = 0$, which shows that (iii) is equivalent to $(\psi^2|1) = 1$, i.e., to $(1/g) \sum_{s \in G} \psi(s)^2 = 1$; since ψ is real-valued, this indeed means that θ is irreducible, cf. th. 5.]

2.4 Decomposition of the regular representation

Notation. For the rest of Ch. 2, the irreducible characters of G are denoted $\chi_1, \ldots, \chi_h$; their degrees are written $n_1, \ldots, n_h$, we have $n_i = \chi_i(1)$, cf. prop. 1.

Let R be the regular representation of G. Recall (cf. 1.2) that it has a basis $(e_t)_{t \in G}$ such that $\rho_s e_t = e_{st}$. If $s \neq 1$, we have $st \neq t$ for all t, which shows that the diagonal terms of the matrix of ρ_s are zero; in particular we have $\mathrm{Tr}(\rho_s) = 0$. On the other hand, for $s = 1$, we have

$$\mathrm{Tr}(\rho_s) = \mathrm{Tr}(1) = \dim(\mathrm{R}) = g.$$

Whence:

Proposition 5. *The character r_{G} of the regular representation is given by the formulas:*

$$r_{\mathrm{G}}(1) = g, \qquad \textit{order of } \mathrm{G},$$
$$r_{\mathrm{G}}(s) = 0 \qquad \textit{if } s \neq 1.$$

Corollary 1. *Every irreducible representation W_i is contained in the regular representation with multiplicity equal to its degree n_i.*

According to th. 4, this number is equal to $\langle r_{\mathrm{G}}, \chi_i \rangle$, and we have

$$\langle r_{\mathrm{G}}, \chi_i \rangle = \frac{1}{g} \sum_{s \in \mathrm{G}} r_{\mathrm{G}}(s^{-1})\chi_i(s) = \frac{1}{g} g \cdot \chi_i(1) = \chi_i(1) = n_i. \qquad \square$$

Corollary 2.

(a) *The degrees n_i satisfy the relation $\sum_{i=1}^{i=h} n_i^2 = g$.*
(b) *If $s \in \mathrm{G}$ is different from 1, we have $\sum_{i=1}^{i=h} n_i \chi_i(s) = 0$.*

By cor. 1, we have $r_{\mathrm{G}}(s) = \sum n_i \chi_i(s)$ for all $s \in \mathrm{G}$. Taking $s = 1$ we obtain (a), and taking $s \neq 1$, we obtain (b). $\square$

Remarks

(1) The above result can be used in determining the irreducible representations of a group G: suppose we have constructed some mutually nonisomorphic irreducible representations of degrees $n_1, \ldots, n_k$; in order that they be *all* the irreducible representations of G (up to isomorphism), it is necessary and sufficient that $n_1^2 + \cdots + n_k^2 = g$.

(2) We will see later (Part II, 6.5) another property of the degrees n_i: they divide the order g of G.

EXERCISE

2.7. Show that each character of G which is zero for all $s \neq 1$ is an *integral* multiple of the character r_{G} of the regular representation.

2.5 Number of irreducible representations

Recall (cf. 2.1) that a function f on G is called a *class function* if $f(tst^{-1}) = f(s)$ for all $s, t \in \mathrm{G}$.

Proposition 6. *Let f be a class function on G, and let $\rho\colon \mathrm{G} \to \mathbf{GL}(\mathrm{V})$ be a linear representation of G. Let ρ_f be the linear mapping of V into itself defined by:*

$$\rho_f = \sum_{t \in \mathrm{G}} f(t)\rho_t.$$

If V is irreducible of degree n and character χ, then ρ_f is a homothety of ratio λ given by:

$$\lambda = \frac{1}{n} \sum_{t \in \mathrm{G}} f(t)\chi(t) = \frac{g}{n}(f|\chi^*).$$

Let us compute $\rho_s^{-1}\rho_f\rho_s$. We have:

$$\rho_s^{-1}\rho_f\rho_s = \sum_{t \in \mathrm{G}} f(t)\rho_s^{-1}\rho_t\rho_s = \sum_{t \in \mathrm{G}} f(t)\rho_{s^{-1}ts}.$$

Putting $u = s^{-1}ts$, this becomes:

$$\rho_s^{-1}\rho_f\rho_s = \sum_{u \in \mathrm{G}} f(sus^{-1})\rho_u = \sum_{u \in \mathrm{G}} f(u)\rho_u = \rho_f.$$

So we have $\rho_f\rho_s = \rho_s\rho_f$. By the second part of prop. 4, this shows that ρ_f is a homothety λ. The trace of λ is $n\lambda$; that of ρ_f is $\sum_{t \in \mathrm{G}} f(t)\mathrm{Tr}(\rho_t) = \sum_{t \in \mathrm{G}} f(t)\chi(t)$. Hence $\lambda = (1/n)\sum_{t \in \mathrm{G}} f(t)\chi(t) = (g/n)(f|\chi^*)$. □

We introduce now the space H of class functions on G; the irreducible characters $\chi_1, \ldots, \chi_h$ belong to H.

Theorem 6. *The characters $\chi_1, \ldots, \chi_h$ form an orthonormal basis of* H.

Theorem 3 shows that the χ_i form an orthonormal system in H. It remains to prove that they *generate* H, and for this it is enough to show that every element of H orthogonal to the χ_i^* is zero. Let f be such an element. For each representation ρ of G, put $\rho_f = \sum_{t \in \mathrm{G}} f(t)\rho_t$. Since f is orthogonal to the χ_i^*, prop. 6 above shows that ρ_f is zero so long as ρ is irreducible; from the direct sum decomposition we conclude that ρ_f is always zero. Applying this to the regular representation R (cf. 2.4) and computing the image of the basis vector e_1 under ρ_f, we have

$$\rho_f e_1 = \sum_{t \in \mathrm{G}} f(t)\rho_t e_1 = \sum_{t \in \mathrm{G}} f(t)e_t.$$

Since ρ_f is zero, we have $\rho_f e_1 = 0$ and the above formula shows that $f(t) = 0$ for all $t \in \mathrm{G}$; hence $f = 0$, and the proof is complete. □

Recall that two elements t and t' of G are said to be conjugate if there exists $s \in \mathrm{G}$ such that $t' = sts^{-1}$; this is an equivalence relation, which partitions G into *classes* (also called *conjugacy classes*).

Theorem 7. *The number of irreducible representations of* G *(up to isomorphism) is equal to the number of classes of* G.

Let $C_1, \ldots, C_k$ be the distinct classes of G. To say that a function f on G is a class function is equivalent to saying that it is constant on each of $C_1, \ldots, C_k$; it is thus determined by its values λ_i on the C_i, and these can be chosen arbitrarily. Consequently, the dimension of the space H of class functions is equal to k. On the other hand, this dimension is, by th. 6, equal to the number of irreducible representations of G (up to isomorphism). The result follows. □

Here is another consequence of th. 6:

Proposition 7. *Let $s \in G$, and let $c(s)$ be the number of elements in the conjugacy class of s.*

(a) *We have* $\sum_{i=1}^{i=h} \chi_i(s)^* \chi_i(s) = g/c(s)$.

(b) *For $t \in G$ not conjugate to s, we have* $\sum_{i=1}^{i=h} \chi_i(s)^* \chi_i(t) = 0$.

(For $s = 1$, this yields cor. 2 to prop. 5.)

Let f_s be the function equal to 1 on the class of s and equal to 0 elsewhere. Since it is a class function, it can, by th. 6, be written

$$f_s = \sum_{i=1}^{i=h} \lambda_i \chi_i, \qquad \text{with } \lambda_i = (f_s|\chi_i) = \frac{c(s)}{g}\chi_i(s)^*.$$

We have then, for each $t \in G$,

$$f_s(t) = \frac{c(s)}{g} \sum_{i=1}^{i=h} \chi_i(s)^* \chi_i(t).$$

This gives (a) if $t = s$, and (b) if t is not conjugate to s. □

Example. Take for G the *group of permutations of three letters*. We have $g = 6$, and there are three classes: the element 1, the three transpositions, and the two cyclic permutations. Let t be a transposition and c a cyclic permutation. We have $t^2 = 1$, $c^3 = 1$, $tc = c^2 t$; whence there are just two characters of degree 1: the unit character χ_1 and the character χ_2 giving the signature of a permutation. Theorem 7 shows that there exists one other irreducible character θ; if n is its degree we must have $1 + 1 + n^2 = 6$, hence $n = 2$. The values of θ can be deduced from the fact that $\chi_1 + \chi_2 + 2\theta$ is the character of the regular representation of G (cf. prop. 5). We thus get the *character table* of G:

	1	t	c
χ_1	1	1	1
χ_2	1	-1	1
θ	2	0	-1

We obtain an irreducible representation with character θ by having G permute the coordinates of elements of $\mathbf{C}^3$ satisfying the equation $x + y + z = 0$ (cf. ex. 2.6c)).

2.6 Canonical decomposition of a representation

Let ρ: G $\rightarrow$ **GL**(V) be a linear representation of G. We are going to define a direct sum decomposition of V which is "coarser" than the decomposition into irreducible representations, but which has the advantage of being *unique*. It is obtained as follows:

Let $\chi_1, \ldots, \chi_h$ be the distinct characters of the irreducible representations $W_1, \ldots, W_h$ of G and $n_1, \ldots, n_h$ their degrees. Let $V = U_1 \oplus \cdots \oplus U_m$ be a decomposition of V into a direct sum of irreducible representations. For $i = 1, \ldots, h$ denote by V_i the direct sum of those of the $U_1, \ldots, U_m$ which are isomorphic to W_i. Clearly we have:

$$V = V_1 \oplus \cdots \oplus V_h.$$

(In other words, we have decomposed V into a direct sum of irreducible representations and *collected together* the isomorphic representations.)

This is the *canonical decomposition* we had in mind. Its properties are as follows:

Theorem 8

(i) *The decomposition* $V = V_1 \oplus \cdots \oplus V_h$ *does not depend on the initially chosen decomposition of* V *into irreducible representations.*

(ii) *The projection* p_i *of* V *onto* V_i *associated with this decomposition is given by the formula*:

$$p_i = \frac{n_i}{g} \sum_{t \in G} \chi_i(t)^* \rho_t.$$

We prove (ii). Assertion (i) will follow because the projections p_i determine the V_i. Put

$$q_i = \frac{n_i}{g} \sum_{t \in G} \chi_i(t)^* \rho_t.$$

Proposition 6 shows that the restriction of q_i to an irreducible representation W with character χ and of degree n is a homothety of ratio $(n_i/n)(\chi_i|\chi)$; it is thus 0 if $\chi \neq \chi_i$ and 1 if $\chi = \chi_i$. In other words q_i is the identity on an irreducible representation isomorphic to W_i, and is zero on the others. In view of the definition of the V_i, it follows that q_i is the identity on V_i and is 0 on V_j for $j \neq i$. If we decompose an element $x \in V$ into its components $x_i \in V_i$:

$$x = x_1 + \cdots + x_h,$$

we have then $q_i(x) = q_i(x_1) + \cdots + q_i(x_h) = x_i$. This means that q_i is equal to the projection p_i of V onto V_i. □

Thus the decomposition of a representation V can be done in two stages. First the canonical decomposition $V_1 \oplus \cdots \oplus V_n$ is determined; this can be done easily using the formulas giving the projections p_i. Next, if needed, one chooses a decomposition of V_i into a direct sum of irreducible representations each isomorphic to W_i:

$$V_i = W_i \oplus \cdots \oplus W_i.$$

This last decomposition can in general be done in an infinity of ways (cf. section 2.7, as well as ex. 2.8 below); it is just as arbitrary as the choice of a basis in a vector space.

EXAMPLE. Take for G the group of two elements $\{1, s\}$ with $s^2 = 1$. This group has two irreducible representations of degree 1, W^+ and W^-, corresponding to $\rho_s = +1$ and $\rho_s = -1$. The canonical decomposition of a representation V is $V = V^+ \oplus V^-$, where V^+ (resp. V^-) consists of the elements $x \in V$ which are symmetric (resp. antisymmetric), i.e., which satisfy $\rho_s x = x$ (resp. $\rho_s x = -x$). The corresponding projections are:

$$p^+ x = \frac{1}{2}(x + \rho_s x), \qquad p^- x = \frac{1}{2}(x - \rho_s x).$$

To decompose V^+ and V^- into irreducible components means simply to decompose these spaces into a direct sum of *lines*.

EXERCISE

2.8. Let H_i be the vector space of linear mappings $h: W_i \to V$ such that $\rho_s h = h\rho_s$ for all $s \in G$. Each $h \in H_i$ maps W_i into V_i.

(a) Show that the dimension of H_i is equal to the number of times that W_i appears in V, i.e., to $\dim V_i/\dim W_i$ [Reduce to the case where $V = W_i$ and use Schur's lemma].

(b) Let G act on $H_i \otimes W_i$ through the tensor product of the trivial representation of G on H_i and the given representation on W_i. Show that the map

$$F: H_i \otimes W_i \to V_i$$

defined by the formula

$$F(\sum h_\alpha \cdot w_\alpha) = \sum h_\alpha(w_\alpha)$$

is an isomorphism of $H_i \otimes W_i$ onto V_i. [Same method.]

(c) Let $(h_1, \ldots, h_k)$ be a basis of H_i and form the direct sum $W_i \oplus \cdots \oplus W_i$ of k copies of W_i. The system $(h_1, \ldots, h_k)$ defines in an obvious way a linear mapping h of $W_i \oplus \cdots \oplus W_i$ into V_i; show that it is an isomorphism of representations and that each isomorphism is thus obtainable [apply (*b*), or argue directly]. In particular, *to decompose* V_i *into a direct sum of representations isomorphic to* W_i *amounts to choosing a basis for* H_i.

2.7 Explicit decomposition of a representation

Keep the notation of the preceding section, and let

$$V = V_1 \oplus \cdots \oplus V_h$$

be the *canonical decomposition* of the given representation. We have seen how one can determine the ith component V_i by means of the corresponding projection (th. 8). We now give a method for explicitly constructing a *decomposition* of V_i into a direct sum of subrepresentations isomorphic to W_i. Let W_i be given in matrix form $(r_{\alpha\beta}(s))$ with respect to a basis $(e_1, \ldots, e_n)$; we have $\chi_i(s) = \sum_\alpha r_{\alpha\alpha}(s)$ and $n = n_i = \dim W_i$. For each pair of integers α, β taken from 1 to n, let $p_{\alpha\beta}$ denote the linear map of V into V defined by

$$(*) \qquad p_{\alpha\beta} = \frac{n}{g} \sum_{t \in G} r_{\beta\alpha}(t^{-1})\rho_t.$$

Proposition 8

(a) *The map $p_{\alpha\alpha}$ is a projection; it is zero on the V_j, $j \neq i$. Its image $V_{i,\alpha}$ is contained in V_i, and V_i is the direct sum of the $V_{i,\alpha}$ for $1 \leqslant \alpha \leqslant n$. We have $p_i = \sum_\alpha p_{\alpha\alpha}$.*

(b) *The linear map $p_{\alpha\beta}$ is zero on the V_j, $j \neq i$, as well as on the $V_{i,\gamma}$ for $\gamma \neq \beta$; it defines an isomorphism from $V_{i,\beta}$ onto $V_{i,\alpha}$.*

(c) *Let x_1 be an element $\neq 0$ of $V_{i,1}$ and let $x_\alpha = p_{\alpha 1}(x_1) \in V_{i,\alpha}$. The x_α are linearly independent and generate a vector subspace $W(x_1)$ stable under G and of dimension n. For each $s \in G$, we have*

$$\rho_s(x_\alpha) = \sum_\beta r_{\beta\alpha}(s) x_\beta$$

(in particular, $W(x_1)$ is isomorphic to W_i).

(d) *If $(x_1^{(1)}, \ldots, x_1^{(m)})$ is a basis of $V_{i,1}$, the representation V_i is the direct sum of the subrepresentations $W(x_1^{(1)}), \ldots, W(x_1^{(m)})$ defined in c).*

(Thus the choice of a basis of $V_{i,1}$ gives a decomposition of V_i into a direct sum of representations isomorphic to W_i.)

We observe first that the formula $(*)$ above allows us to define the $p_{\alpha\beta}$ in arbitrary representations of G, and in particular in the irreducible representations W_j. For W_i, we have

$$p_{\alpha\beta}(e_\gamma) = \frac{n}{g} \sum_{t \in G} r_{\beta\alpha}(t^{-1})\rho_t(e_\gamma) = \frac{n}{g} \sum_\delta \sum_{t \in G} r_{\beta\alpha}(t^{-1}) r_{\delta\gamma}(t) e_\delta.$$

By cor. 3 to prop. 4 we have then

$$p_{\alpha\beta}(e_\gamma) = \begin{cases} e_\alpha & \text{if } \gamma = \beta \\ 0 & \text{otherwise}. \end{cases}$$

We get from this the fact that $\sum_{\alpha} p_{\alpha\alpha}$ is the identity map of W_i, and the formulas

$$p_{\alpha\beta} \circ p_{\gamma\delta} = \begin{cases} p_{\alpha\delta} & \text{if } \beta = \gamma \\ 0 & \text{otherwise} \end{cases}$$

$$\rho_s \circ p_{\alpha\gamma} = \sum_{\beta} r_{\beta\alpha}(s) p_{\beta\gamma}.$$

For W_j with $j \neq i$, we use cor. 2 to prop. 4 and the same argument to show that all the $p_{\alpha\beta}$ are *zero.*

Having done this, we decompose V into a direct sum of subrepresentations isomorphic to W_i and apply the preceding to each of these representations. Assertions (a) and (b) follow; moreover, the above formulas remain valid in V. Under the hypothesis of (c), we have then

$$\rho_s(x_\alpha) = \rho_s \circ p_{\alpha 1}(x_1) = \sum_{\beta} r_{\beta\alpha}(s) p_{\beta 1}(x_1) = \sum_{\beta} r_{\beta\alpha}(s) x_\beta,$$

which proves (c). Finally (d) follows from (a), (b), and (c). □

Exercises

2.9. Let H_i be the space of linear maps h: $W_i \to V$ such that $h \circ \rho_s = \rho_s \circ h$, cf. ex. 2.8. Show that the map $h \mapsto h(e_\alpha)$ is an isomorphism of H_i onto $V_{i,\alpha}$.

2.10. Let $x \in V_i$, and let $V(x)$ be the smallest subrepresentation of V containing x. Let x_1^α be the image of x under $p_{1\alpha}$; show that $V(x)$ is the sum of the representations $W(x_1^\alpha)$, $\alpha = 1, \ldots, n$. Deduce from this that $V(x)$ is the direct sum of at most n subrepresentations isomorphic to W_i.

CHAPTER 3
Subgroups, products, induced representations

All the groups considered below are assumed to be finite.

3.1 Abelian subgroups

Let G be a group. One says that G is *abelian* (or *commutative*) if $st = ts$ for all $s, t \in G$. This amounts to saying that each conjugacy class of G consists of a single element, also that each function on G is a class function. The linear representations of such a group are particularly simple:

Theorem 9. *The following properties are equivalent*:

(i) G *is abelian.*
(ii) *All the irreducible representations of* G *have degree* 1.

Let g be the order of G, and let $(n_1, \ldots, n_h)$ be the degrees of the distinct irreducible representations of G; we know, cf. Ch. 2, that h is the number of classes of G, and that $g = n_1^2 + \cdots + n_h^2$. Hence g is equal to h if and only if all the n_i are equal to 1, which proves the theorem. □

Corollary. *Let* A *be an abelian subgroup of* G, *let a be its order and let g be that of* G. *Each irreducible representation of* G *has degree* $\leqslant g/a$.

(The quotient g/a is the *index* of A in G.)

Let ρ: G $\to$ **GL**(V) be an irreducible representation of G. Through *restriction* to the subgroup A, it defines a representation ρ_A : A $\to$ **GL**(V) of A. Let W $\subset$ V be an irreducible subrepresentation of ρ_A; by th. 9, we have dim(W) $= 1$. Let V′ be the vector subspace of V generated by the images ρ_sW of W, s ranging over G. It is clear that V′ is stable under G; since ρ is irreducible, we thus have V′ = V. But, for $s \in$ G and $t \in$ A we have

$$\rho_{st} W = \rho_s \rho_t W = \rho_s W.$$

It follows that the number of distinct $\rho_s W$ is at most equal to g/a, hence the desired inequality $\dim(V) \leqslant g/a$, since V is the sum of the $\rho_s W$. □

EXAMPLE. A dihedral group contains a cyclic subgroup of index 2. Its irreducible representations thus have degree 1 or 2; we will determine them later (5.3).

EXERCISES

3.1. Show directly, using Schur's lemma, that each irreducible representation of an abelian group, finite or not, has degree 1.

3.2. Let ρ be an irreducible representation of G of degree n and character χ; let C be the *center* of G (i.e., the set of $s \in G$ such that $st = ts$ for all $t \in G$), and let c be its order.

(a) Show that ρ_s is a homothety for each $s \in C$. [Use Schur's lemma.] Deduce from this that $|\chi(s)| = n$ for all $s \in C$.

(b) Prove the inequality $n^2 \leqslant g/c$. [Use the formula $\sum_{s \in G} |\chi(s)|^2 = g$, combined with (a).]

(c) Show that, if ρ is *faithful* (i.e., $\rho_s \neq 1$ for $s \neq 1$), the group C is cyclic.

3.3. Let G be an abelian group of order g, and let $\hat{G}$ be the set of irreducible characters of G. If χ_1, χ_2 belong to $\hat{G}$, the same is true of their product $\chi_1 \chi_2$. Show that this makes $\hat{G}$ an abelian group of order g; the group $\hat{G}$ is called the dual of the group G. For $x \in G$ the mapping $\chi \mapsto \chi(x)$ is an irreducible character of $\hat{G}$ and so an element of the dual $\hat{\hat{G}}$ of $\hat{G}$. Show that the map of G into $\hat{\hat{G}}$ thus obtained is an injective homomorphism; conclude (by comparing the orders of the two groups) that it is an *isomorphism*.

3.2 Product of two groups

Let G_1 and G_2 be two groups, and let $G_1 \times G_2$ be their *product*, that is, the set of pairs (s_1, s_2), with $s_1 \in G_1$ and $s_2 \in G_2$.

Putting

$$(s_1, s_2) \cdot (t_1, t_2) = (s_1 t_1, s_2 t_2),$$

we define a group structure on $G_1 \times G_2$; endowed with this structure, $G_1 \times G_2$ is called the *group product* of G_1 and G_2. If G_1 has order g_1 and G_2 has order g_2, $G_1 \times G_2$ has order $g = g_1 g_2$. The group G_1 can be identified with the subgroup of $G_1 \times G_2$ consisting of elements $(s_1, 1)$, where s_1 ranges over G_1; similarly, G_2 can be identified with a subgroup of $G_1 \times G_2$. With these identifications, each element of G_1 *commutes* with each element of G_2.

Conversely, let G be a group containing G_1 and G_2 as subgroups, and suppose the following two conditions are satisfied:

(i) *Each $s \in G$ can be written uniquely in the form $s = s_1 s_2$ with $s_1 \in G_1$ and $s_2 \in G_2$.*

(ii) *For $s_1 \in G_1$ and $s_2 \in G_2$, we have $s_1 s_2 = s_2 s_1$.*

The product of two elements $s = s_1 s_2$, $t = t_1 t_2$ can then be written

$$st = s_1 s_2 t_1 t_2 = (s_1 t_1)(s_2 t_2).$$

It follows that, if we let $(s_1, s_2) \in \mathrm{G}_1 \times \mathrm{G}_2$ correspond to the element $s_1 s_2$ of G, we obtain an *isomorphism of* $\mathrm{G}_1 \times \mathrm{G}_2$ *onto* G. In this case, we also say that G is the *product* (or the *direct product*) of its subgroups G_1 and G_2, and we identify it with $\mathrm{G}_1 \times \mathrm{G}_2$.

Now let $\rho^1 \colon \mathrm{G}_1 \to \mathbf{GL}(\mathrm{V}_1)$ and $\rho^2 \colon \mathrm{G}_2 \to \mathbf{GL}(\mathrm{V}_2)$ be linear representations of G_1 and G_2 respectively. We define a linear representation $\rho^1 \otimes \rho^2$ of $\mathrm{G}_1 \times \mathrm{G}_2$ into $\mathrm{V}_1 \otimes \mathrm{V}_2$ by a procedure analogous to 1.5 by setting

$$(\rho^1 \otimes \rho^2)(s_1, s_2) = \rho^1(s_1) \otimes \rho^2(s_2).$$

This representation is called the *tensor product* of the representations ρ^1 and ρ^2. If χ_i is the character of ρ_i $(i = 1, 2)$, the character χ of $\rho^1 \otimes \rho^2$ is given by:

$$\chi(s_1, s_2) = \chi_1(s_1) \cdot \chi_2(s_2).$$

When G_1 and G_2 are equal to the same group G, the representation $\rho^1 \otimes \rho^2$ defined above is a representation of $\mathrm{G} \times \mathrm{G}$. When restricted to the *diagonal* subgroup of $\mathrm{G} \times \mathrm{G}$ (consisting of (s, s), where s ranges over G), it gives the representation of G denoted $\rho^1 \otimes \rho^2$ in 1.5; in spite of the identity of notations, it is important to distinguish these two representations.

Theorem 10

(i) *If ρ^1 and ρ^2 are irreducible, $\rho^1 \otimes \rho^2$ is an irreducible representation of $\mathrm{G}_1 \times \mathrm{G}_2$.*

(ii) *Each irreducible representation of $\mathrm{G}_1 \times \mathrm{G}_2$ is isomorphic to a representation $\rho^1 \otimes \rho^2$, where ρ^i is an irreducible representation of G_i $(i = 1, 2)$.*

If ρ^1 and ρ^2 are irreducible, we have (cf. 2.3):

$$\frac{1}{g_1} \sum_{s_1} |\chi_1(s_1)|^2 = 1, \qquad \frac{1}{g_2} \sum_{s_2} |\chi_2(s_2)|^2 = 1.$$

By multiplication, this gives:

$$\frac{1}{g} \sum_{s_1, s_2} |\chi(s_1, s_2)|^2 = 1$$

which shows that $\rho^1 \otimes \rho^2$ is irreducible (th. 5). In order to prove (ii), it suffices to show that each class function f on $\mathrm{G}_1 \times \mathrm{G}_2$, which is orthogonal to the characters of the form $\chi_1(s_1)\chi_2(s_2)$, is zero. Suppose then that we have:

$$\sum_{s_1, s_2} f(s_1, s_2)\chi_1(s_1)^* \chi_2(s_2)^* = 0.$$

Fixing χ_2 and putting $g(s_1) = \sum_{s_2} f(s_1, s_2)\chi_2(s_2)^*$ we have:

$$\sum_{s_1} g(s_1)\chi_1(s_1)^* = 0 \quad \text{for all } \chi_1.$$

Since g is a class function, this implies $g = 0$, and, since the same is true for each χ_2, we conclude by the same argument that $f(s_1, s_2) = 0$. □

[It is also possible to prove (ii) by computing the sum of the squares of the degrees of the representations $\rho^1 \otimes \rho^2$, and applying 2.4.]

The above theorem completely reduces the study of representations of $G_1 \times G_2$ to that of representations of G_1 and of representations of G_2.

3.3 Induced representations

Left cosets of a subgroup

Recall the following definition: Let H be a subgroup of a group G. For $s \in G$, we denote by sH the set of products st with $t \in H$, and say that sH is the *left coset* of H containing s. Two elements s, s' of G are said to be *congruent modulo* H if they belong to the same left coset, i.e., if $s^{-1}s'$ belongs to H; we write then $s' \equiv s \pmod{H}$. The set of left cosets of H is denoted by G/H; it is a partition of G. If G has g elements and H has h elements, G/H has g/h elements; the integer g/h is the *index* of H in G and is denoted by (G:H).

If we choose an element from each left coset of H, we obtain a subset R of G called a system of representatives of G/H; each s in G can be written uniquely $s = rt$, with $r \in R$ and $t \in H$.

Definition of induced representations

Let ρ: G → **GL**(V) be a linear representation of G, and let ρ_H be its restriction to H. Let W be a subrepresentation of ρ_H, that is, a vector subspace of V stable under the ρ_t, $t \in H$. Denote by θ: H → **GL**(W) the representation of H in W thus defined. Let $s \in G$; the vector space $\rho_s W$ depends only on the left coset sH of s; indeed, if we replace s by st, with $t \in H$, we have $\rho_{st}W = \rho_s\rho_t W = \rho_s W$ since $\rho_t W = W$. If σ is a left coset of H, we can thus define a subspace W_σ of V to be $\rho_s W$ for any $s \in \sigma$. It is clear that the W_σ are permuted among themselves by the ρ_s, $s \in G$. Their sum $\sum_{\sigma \in G/H} W_\sigma$ is thus a subrepresentation of V.

Definition. We say that the representation ρ of G in V is *induced* by the representation θ of H in W if V is equal to the sum of the W_σ ($\sigma \in G/H$) and if this sum is direct (that is, if $V = \bigoplus_{\sigma \in G/H} W_\sigma$).

We can reformulate this condition in several ways:

(i) Each $x \in V$ can be written uniquely as $\sum_{\sigma \in G/H} x_\sigma$, with $x_\sigma \in W_\sigma$ for each σ.

(ii) If R is a system of representatives of G/H, the vector space V is the *direct sum* of the $\rho_r W$, with $r \in R$.

In particular, we have $\dim(V) = \sum_{r \in R} \dim(\rho_r W) = (G\colon H) \cdot \dim(W)$.

EXAMPLES 1. Take for V the regular representation of G; the space V has a basis $(e_t)_{t \in G}$ such that $\rho_s e_t = e_{st}$ for $s \in G$, $t \in G$. Let W be the subspace of V with basis $(e_t)_{t \in H}$. The representation θ of H in W is the *regular representation* of H, and it is clear that ρ is induced by θ.

2. Take for V a vector space having a basis (e_σ) indexed by the elements σ of G/H and define a representation ρ of G in V by $\rho_s e_\sigma = e_{s\sigma}$ for $s \in G$ and $\sigma \in G/H$ (this formula makes sense, because, if σ is a left coset of H, so is $s\sigma$). We thus obtain a representation of G which is the *permutation representation* of G associated with G/H [cf. 1.2, example (*c*)]. The vector e_H corresponding to the coset H is invariant under H; the representation of H in the subspace $\mathbf{C}e_H$ is thus the *unit representation* of H, and it is clear that this representation induces the representation ρ of G in V.

3. If ρ_1 is induced by θ_1 and if ρ_2 is induced by θ_2, then $\rho_1 \oplus \rho_2$ is induced by $\theta_1 \oplus \theta_2$.

4. If (V, ρ) is induced by (W, θ), and if W_1 is a stable subspace of W, the subspace $V_1 = \sum_{r \in R} \rho_r W_1$ of V is stable under G, and the representation of G in V_1 is induced by the representation of H in W_1.

5. If ρ is induced by θ, if ρ' is a representation of G, and if ρ'_H is the restriction of ρ' to H, then $\rho \otimes \rho'$ is induced by $\theta \otimes \rho'_H$.

Existence and uniqueness of induced representations

Lemma 1. *Suppose that* (V, ρ) *is induced by* (W, θ). *Let* $\rho'\colon G \to \mathbf{GL}(V')$ *be a linear representation of* G, *and let* $f\colon W \to V'$ *be a linear map such that* $f(\theta_t w) = \rho'_t f(w)$ *for all* $t \in H$ *and* $w \in W$. *Then there exists a unique linear map* $F\colon V \to V'$ *which extends* f *and satisfies* $F \circ \rho_s = \rho'_s \circ F$ *for all* $s \in G$.

If F satisfies these conditions, and if $x \in \rho_s W$, we have $\rho_s^{-1} x \in W$; hence

$$F(x) = F(\rho_s \rho_s^{-1} x) = \rho'_s F(\rho_s^{-1} x) = \rho'_s f(\rho_s^{-1} x).$$

This formula determines F on $\rho_s W$, and so on V, since V is the sum of the $\rho_s W$. This proves the uniqueness of F.

Now let $x \in W_\sigma$, and choose $s \in \sigma$; we *define* $F(x)$ by the formula $F(x) = \rho'_s f(\rho_s^{-1} x)$ as above. This definition does not depend on the choice of s in σ; indeed, if we replace s by st, with $t \in H$, we have

$$\rho'_{st} f(\rho_{st}^{-1} x) = \rho'_s \rho'_t f(\theta_t^{-1} \rho_s^{-1} x) = \rho'_s(\theta_t \theta_t^{-1} \rho_s^{-1} x) = \rho'_s f(\rho_s^{-1} x).$$

Since V is the direct sum of the W_σ, there exists a unique linear map

$F: V \to V'$ which extends the partial mappings thus defined on the W_σ. It is easily checked that $F \circ \rho_s = \rho'_s \circ F$ for all $s \in G$. □

Theorem 11. *Let* (W, θ) *be a linear representation of* H. *There exists a linear representation* (V, ρ) *of* G *which is induced by* (W, θ), *and it is unique up to isomorphism.*

Let us first prove the existence of the induced representation ρ. In view of example 3, above, we may assume that θ is irreducible. In this case, θ is isomorphic to a subrepresentation of the regular representation of H, which can be induced to the regular representation of G (cf. example 1). Applying example 4, we conclude that θ itself can be induced.

It remains to prove the uniqueness of ρ up to isomorphism. Let (V, ρ) and (V', ρ') be two representations induced by (W, θ). Applying Lemma 1 to the injection of W into V′, we see that there exists a linear map $F: V \to V'$ which is the identity on W and satisfies $F \circ \rho_s = \rho'_s \circ F$ for all $s \in G$. Consequently the image of F contains all the $\rho'_s W$, and thus is equal to V′. Since V′ and V have the same dimension $(G: H) \cdot \dim(W)$, we see that F is an *isomorphism*, which proves the theorem. (For a more natural proof of Theorem 11, see 7.1.) □

Character of an induced representation

Suppose (V, ρ) is induced by (W, θ) and let χ_ρ and χ_θ be the corresponding characters of G and of H. Since (W, θ) determines (V, ρ) up to isomorphism, we ought to be able to compute χ_ρ from χ_θ. The following theorem tells how:

Theorem 12. *Let h be the order of* H *and let* R *be a system of representatives of* G/H. *For each* $u \in G$, *we have*

$$\chi_\rho(u) = \sum_{\substack{r \in R \\ r^{-1}ur \in H}} \chi_\theta(r^{-1}ur) = \frac{1}{h} \sum_{\substack{s \in G \\ s^{-1}us \in H}} \chi_\theta(s^{-1}us).$$

(In particular, $\chi_\rho(u)$ is a linear combination of the values of χ_θ on the intersection of H with the conjugacy class of u in G.)

The space V is the direct sum of the $\rho_r W$, $r \in R$. Moreover ρ_u permutes the $\rho_r W$ among themselves. More precisely, if we write ur in the form $r_u t$ with $r_u \in R$ and $t \in H$, we see that ρ_u sends $\rho_r W$ into $\rho_{r_u} W$. To determine $\chi_\rho(u) = \mathrm{Tr}_V(\rho_u)$, we can use a basis of V which is a union of bases of the $\rho_r W$. The indices r such that $r_u \neq r$ give *zero* diagonal terms; the others give the trace of ρ_u on the $\rho_r W$. We thus obtain:

$$\chi_\rho(u) = \sum_{r \in R_u} \mathrm{Tr}_{\rho_r W}(\rho_{u,r}),$$

where R_u denotes the set of $r \in R$ such that $r_u = r$ and $\rho_{u,r}$ is the restriction of ρ_u to $\rho_r W$. Observe that r belongs to R_u if and only if ur can be written rt, with $t \in H$, i.e., if $r^{-1}ur$ belongs to H.

It remains to compute $\mathrm{Tr}_{\rho_r W}(\rho_{u,r})$, for $r \in R_u$. To do this, note that ρ_r defines an isomorphism of W onto $\rho_r W$, and that we have

$$\rho_r \circ \theta_t = \rho_{u,r} \circ \rho_r, \quad \text{with } t = r^{-1}ur \in H.$$

The trace of $\rho_{u,r}$ is thus equal to that of θ_t, that is, to $\chi_\theta(t) = \chi_\theta(r^{-1}ur)$. We indeed obtain:

$$\chi_\rho(u) = \sum_{r \in R_u} \chi_\theta(r^{-1}ur).$$

The second formula given for $\chi_\rho(u)$ follows from the first by noting that all elements s of G in the left coset rH ($r \in R_u$) satisfy $\chi_\theta(s^{-1}us) = \chi_\theta(r^{-1}ur)$. □

The reader will find other properties of induced representations in part II. Notably:

(i) The *Frobenius reciprocity formula*

$$(f_H | \chi_\theta)_H = (f | \chi_\rho)_G$$

where f is a class function of G, and f_H is its restriction to H, and the scalar products are calculated on H and G respectively.

(ii) *Mackey's criterion*, which tells us when an induced representation is irreducible.

(iii) *Artin's theorem* (resp. *Brauer's theorem*), which says that each character of a group G is a linear combination with rational (resp. integral) coefficients of characters of representations induced from cyclic subgroups (resp. from "elementary" subgroups) of G.

Exercises

3.4. Show that each irreducible representation of G is contained in a representation induced by an irreducible representation of H. [Use the fact that an irreducible representation is contained in the regular representation.] Obtain from this another proof of the cor. to th. 9.

3.5. Let (W, θ) be a linear representation of H. Let V be the vector space of functions $f\colon G \to W$ such that $f(tu) = \theta_t f(u)$ for $u \in G$, $t \in H$. Let ρ be the representation of G in V defined by $(\rho_s f)(u) = f(us)$ for $s, u \in G$. For $w \in W$ let $f_w \in V$ be defined by $f_w(t) = \theta_t w$ for $t \in H$ and $f_w(s) = 0$ for $s \notin H$. Show that $w \mapsto f_w$ is an isomorphism of W onto the subspace W_0 of V consisting of functions which vanish off H. Show that, if we identify W and W_0 in this way, the representation (V, ρ) is induced by the representation (W, θ).

3.6. Suppose that G is the direct product of two subgroups H and K (cf. 3.2). Let ρ be a representation of G induced by a representation θ of H. Show that ρ is isomorphic to $\theta \otimes r_K$, where r_K denotes the regular representation of K.

CHAPTER 4

Compact groups

The purpose of this chapter is to indicate how the preceding results carry over to arbitrary *compact* groups (not necessarily finite); for the proofs, see [1], [4], [6] cited in the bibliography.

None of the results below will be used in the sequel, aside from examples 5.2, 5.5, and 5.6.

4.1 Compact groups

A *topological group* G is a group endowed with a topology such that the product $s \cdot t$ and the inverse s^{-1} are continuous. Such a group is said to be *compact* if its topology is that of a compact space, that is, satisfies the Borel–Lebesgue theorem. For example, the group of *rotations* around a point in euclidean space of dimension 2 (or 3, ...) has a natural topology which makes it into a compact group; its *closed subgroups* are also compact groups.

As examples of *noncompact* groups, we mention the group of *translations* $x \mapsto x + a$, and the group of linear mappings preserving the quadratic form $x^2 + y^2 + z^2 - t^2$ (the "Lorentz group"). The linear representations of these groups have completely different properties from those in the compact case.

4.2 Invariant measure on a compact group

In the study of linear representations of a finite group G of order g, we have used a great deal the operation of *averaging* over G, i.e., attaching to a function f on G the element $(1/g) \sum_{t \in \mathrm{G}} f(t)$ (the values of f could be either complex numbers or, more generally, elements of a vector space). An analogous operation exists for compact groups; of course, instead of a finite sum, we have an integral $\int_G f(t)\,dt$ with respect to a measure dt.

More precisely, one proves the *existence and uniqueness* of a measure dt carried by G and enjoying the following two properties:

(i) $\int_G f(t)\,dt = \int_G f(ts)\,dt$ *for each continuous function* f *and each* $s \in G$ (*invariance of* dt *under right translation*).

(ii) $\int_G dt = 1$ (*the total mass of* dt *is equal to* 1).

One shows moreover that dt is *invariant under left translation*, i.e.:

$$\text{(i}') \qquad \int_G f(t)\,dt = \int_G f(st)\,dt.$$

The measure dt is called the *invariant measure* (or *Haar measure*) of the group G. We give two examples (see also Ch. 5):

(1) If G is finite of order g, the measure dt is obtained by assigning to each element $t \in G$ a mass equal to $1/g$.

(2) If G is the group C_∞ of rotations in the plane, and if we represent the elements $t \in G$ in the form $t = e^{i\alpha}$ (α taken modulo 2π), the invariant measure is $(1/2\pi)d\alpha$; the factor $1/2\pi$ is used to insure condition (ii).

4.3 Linear representations of compact groups

Let G be a compact group and let V be a vector space of finite dimension over the field of complex numbers. A linear representations of G in V is a homomorphism $\rho: G \to \mathbf{GL}(V)$ which is *continuous*; this condition is equivalent to saying that $\rho_s x$ is a continuous function of the two variables $s \in G$, $x \in V$. One defines similarly linear representations of G in a *Hilbert space*; one proves, moreover, that such a representation is isomorphic to a (Hilbert) *direct sum of unitary representations of finite dimension*, which allows one to restrict attention to the latter.

Most of the properties of representations of finite groups carry over to representations of compact groups; one just replaces the expressions "$(1/g)\sum_{t \in G} f(t)$" by "$\int_G f(t)\,dt$". For example, the *scalar product* $(\phi|\psi)$ of two functions ϕ and ψ is

$$(\phi|\psi) = \int_G \phi(t)\psi(t)^*\,dt.$$

More precisely:

(a) Theorems 1, 2, 3, 4, and 5 carry over without change, as well as their proofs. The same holds for propositions 1, 2, 3, and 4.

(b) In 2.4, it is necessary to define the regular representation R as the Hilbert space of square integrable functions on G with group action $(\rho_s f)(t) = f(s^{-1}t)$. If G is not finite, this representation is of infinite dimension, and it is no longer possible to speak of its character, so proposition 5 no longer makes sense. Nevertheless, it is still true that each irreducible representation is contained in R with multiplicity equal to its degree.

(c) Proposition 6 and th. 6 carry over without change (in th. 6, take for H the Hilbert space of square integrable functions on G).

(d) Theorem 7 is true (but uninteresting) when G is not finite: there are infinitely many classes, and infinitely many irreducible representations.

(e) Theorem 8 and prop. 8 carry over without change, as well as their proofs. The projections p_i of the canonical decomposition (th. 8) are given by the formulas

$$p_i x = n_i \int_G \chi_i(t)^* \rho_t x \, dt.$$

(f) Theorems 9 and 10 carry over without change, as well as their proofs. Note, with respect to th. 10, that the invariant measure of the product $G_1 \times G_2$ is the product $ds_1 \, ds_2$ of the invariant measures of the groups G_1 and G_2.

(g) So long as H is a closed subgroup of *finite index* in G, the notion of a representation of G *induced* by a representation of H, defined as in 3.3, and th. 11 and 12, remain valid. When the index of H is infinite, the representation induced by (W, θ) is defined as the Hilbert space of square integrable functions f on G, with values in W, such that $f(tu) = \theta_t f(u)$ for each $t \in H$, and G acts on this space by $\rho_s f(u) = f(us)$, cf. ex. 3.5.

CHAPTER 5
Examples

5.1 The cyclic group C_n

This is the group of order n consisting of the powers $1, r, \ldots, r^{n-1}$ of an element r such that $r^n = 1$. It can be realized as the group of rotations through angles $2k\pi/n$ around an axis. It is an abelian group.

According to th. 9, the irreducible representations of C_n are of degree 1. Such a representation associates with r a complex number $\chi(r) = w$, and with r^k the number $\chi(r^k) = w^k$; since $r^n = 1$, we have $w^n = 1$, that is, $w = e^{2\pi i h/n}$, with $h = 0, 1, \ldots, n-1$. We thus obtain n *irreducible representations of degree* 1 whose characters $\chi_0, \chi_1, \ldots, \chi_{n-1}$ are given by

$$\chi_h(r^k) = e^{2\pi i hk/n}.$$

We have $\chi_h \cdot \chi_{h'} = \chi_{h+h'}$, with the convention that $\chi_{h+h'} = \chi_{h+h'-n}$ if $h + h' \geqslant n$ (in other words, the index h of χ_h is taken *modulo* n).

For $n = 3$, for example, the *character table* is the following:

	1	r	r^2
χ_0	1	1	1
χ_1	1	w	w^2
χ_2	1	w^2	w

where

$$w = e^{2\pi i/3} = -\frac{1}{2} + i\frac{\sqrt{3}}{2}.$$

We have

$$\chi_0 \cdot \chi_i = \chi_i, \chi_1 \cdot \chi_1 = \chi_2, \chi_2 \cdot \chi_2 = \chi_1 \text{ and } \chi_1 \cdot \chi_2 = \chi_0.$$

5.2 The group C_∞

This is the group of *rotations* of the plane. If we denote by r_α the rotation through an angle α (determined modulo 2π), the *invariant measure* on C_∞ is $(1/2\pi)\,d\alpha$ (cf. 4.2).

The irreducible representations of C_∞ are of degree 1. They are given by:

$$\chi_n(r_\alpha) = e^{in\alpha} \quad (n \text{ an arbitrary integer}).$$

The orthogonality relations give here the well known formulas:

$$\frac{1}{2\pi}\int_0^{2\pi} e^{-in\alpha}\cdot e^{im\alpha}\,d\alpha = \delta_{nm},$$

and th. 6 gives the expansion of a periodic function as a Fourier series.

5.3 The dihedral group D_n

This is the group of rotations and reflections of the plane which preserve a regular polygon with n vertices. It contains n rotations, which form a subgroup isomorphic to C_n, and n reflections. Its order is $2n$. If we denote by r the rotation through an angle $2\pi/n$ and if s is any one of the reflections, we have:

$$r^n = 1, \qquad s^2 = 1, \qquad srs = r^{-1}.$$

Each element of D_n can be written uniquely, either in the form r^k, with $0 \leqslant k \leqslant n-1$ (if it belongs to C_n), or in the form sr^k, with $0 \leqslant k \leqslant n-1$ (if it does not belong to C_n). Observe that the relation $srs = r^{-1}$ implies $sr^ks = r^{-k}$, whence $(sr^k)^2 = 1$.

Realization of D_n *as a group of rigid motions of 3-space*

There are several such:

(a) The usual realization (the one traditionally denoted D_n cf. Eyring [5]). One takes for rotations the rotations around the axis Oz, and for reflections, the reflections through n lines of the plane Oxy, these lines forming angles which are multiples of π/n.

(b) The realization by means of the group C_{nv} (notation of Eyring [5]): instead of the reflections with respect to the *lines* of Oxy, one takes reflections with respect to *planes* containing the axis Oz.

(c) The group D_{2n} can also be realized as the group D_{nd} (notation of Eyring [5]).

Irreducible representations of the group D_n *(n even* $\geqslant 2$*)*

First, there are 4 representations of degree 1, obtained by letting ± 1 correspond to r and s in all possible ways. Their characters $\psi_1, \psi_2, \psi_3, \psi_4$ are given by the following table:

	r^k	sr^k
ψ_1	1	1
ψ_2	1	-1
ψ_3	$(-1)^k$	$(-1)^k$
ψ_4	$(-1)^k$	$(-1)^{k+1}$

Next we consider representations of degree 2. Put $w = e^{2\pi i/n}$ and let h be an arbitrary integer. We define a representation ρ^h of D_n by setting:

$$\rho^h(r^k) = \begin{pmatrix} w^{hk} & 0 \\ 0 & w^{-hk} \end{pmatrix}, \qquad \rho^h(sr^k) = \begin{pmatrix} 0 & w^{-hk} \\ w^{hk} & 0 \end{pmatrix}.$$

A direct calculation shows that this is indeed a representation. This representation is *induced* (in the sense of 3.3) by the representation of C_n with character χ_h (5.1). It depends only on the residue class of h modulo n; moreover ρ^h and ρ^{n-h} are isomorphic. Hence we may assume $0 \leqslant h \leqslant n/2$. The extreme cases $h = 0$ and $h = n/2$ are uninteresting: the corresponding representations are reducible, with characters $\psi_1 + \psi_2$ and $\psi_3 + \psi_4$ respectively. On the other hand, for $0 < h < n/2$, the representation ρ^h is *irreducible*: since $w^h \neq w^{-h}$, the only lines stable under $\rho^h(r)$ are the coordinate axes, and these are not stable under $\rho^h(s)$. The same argument shows that these representations are pairwise nonisomorphic. The corresponding characters χ^h are given by:

$$\chi_h(r^k) = w^{hk} + w^{-hk} = 2\cos\frac{2\pi hk}{n}$$

$$\chi_h(sr^k) = 0.$$

The irreducible representations of degree 1 and 2 constructed above are the *only irreducible representations* of D_n (up to isomorphism). Indeed, the sum of the squares of their degrees is equal to $4 \times 1 + ((n/2) - 1) \times 4 = 2n$, which is the order of D_n.

Example. The group D_6 has 4 representations of degree 1, with characters $\psi_1, \psi_2, \psi_3, \psi_4$ and 2 irreducible representations of degree 2, with characters χ_1 and χ_2.

Irreducible representations of the group D_n *(n odd)*

There are only two representations of degree 1, and their characters ψ_1 and ψ_2 are given by the table:

	r^k	sr^k
ψ_1	1	1
ψ_2	1	-1

The representations ρ^h of degree 2 are defined by the same formulas as in the case where n is even. Those corresponding to $0 < h < n/2$ are irreducible and pairwise nonisomorphic (observe that, since n is odd, the condition $h < n/2$ can also be written $h \leqslant (n-1)/2$). The formulas giving their characters are the same.

These representations are the *only* ones. Indeed, the sum of the squares of their degrees is equal to $2 \times 1 + \frac{1}{2}(n-1) \times 4 = 2n$, and this is the order of D_n.

EXERCISES

5.1. Show that in D_n, n even (resp. odd), the reflections form two conjugacy classes (resp. one), and that the elements of C_n form $(n/2)+1$ classes (resp. $(n+1)/2$ classes). Obtain from this the number of classes of D_n and check that it coincides with the number of irreducible characters.

5.2. Show that $\chi_h \cdot \chi_{h'} = \chi_{h+h'} + \chi_{h-h'}$. In particular, we have

$$\chi_h \cdot \chi_h = \chi_{2h} + \chi_0 = \chi_{2h} + \psi_1 + \psi_2.$$

Show that ψ_2 is the character of the alternating square of ρ^h, and that $\chi_{2h} + \psi_1$ is the character of its symmetric square (cf. 1.5 and prop. 3).

5.3. Show that the usual realization of D_n as a group of rigid motions in $\mathbf{R}^3$ (Eyring [5]) is reducible and has character $\chi_1 + \psi_2$, and that the realization of D_n as C_{nv} (*loc. cit.*) has $\chi_1 + \psi_1$ for its character.

5.4 The group D_{nh}

This group is the product $D_n \times I$, where I is a group of order 2 consisting of elements $\{1, \iota\}$ with $\iota^2 = 1$. Its order is $4n$. If D_n is realized in the usual way as a group of rotations and reflections of 3-space [cf. 5.3, (a)] then D_{nh} can be realized as the group generated by D_n and the reflection ι through the origin.

According to th. 10, the irreducible representations of D_{nh} are the tensor products of those of D_n and those of I. The group I has just two irreducible representations, both of degree 1. Their characters g and u are given by the table:

	1	ι
g	1	1
u	1	-1

Consequently, D_{nh} *has twice as many irreducible representations as* D_n. More precisely, each irreducible character χ of D_n defines two irreducible characters χ_g and χ_u of D_{nh} as follows:

	x	ιx
χ_g	$\chi(x)$	$\chi(x)$
χ_u	$\chi(x)$	$-\chi(x)$

$(\chi \in D_n)$

For example, the character χ_1 of D_n gives rise to characters χ_{1g} and χ_{1u}:

	r^k	sr^k	ιr^k	ιsr^k
χ_{1g}	$2\cos 2\pi k/n$	0	$2\cos 2\pi k/n$	0
χ_{1u}	$2\cos 2\pi k/n$	0	$-2\cos 2\pi k/n$	0

The same applies to the other characters of D_n.

5.5 The group D_∞

This is the group of rotations and reflections of the plane which preserve the origin. It contains the group C_∞ of rotations r_α; if s is an arbitrary reflection, we have the relations:

$$s^2 = 1, \qquad sr_\alpha s = r_{-\alpha}.$$

Each element of D_∞ can be written uniquely either in the form r_α (if it belongs to C_∞) or in the form sr_α (if it does not belong to C_∞); as a topological space, D_∞ consists of two disjoint circles. The *invariant measure* of D_∞ is the measure $d\alpha/4\pi$. More precisely, the *average* $\int_G f(t)\,dt$ of a function f is given by the formula

$$\int_G f(t)\,dt = \frac{1}{4\pi}\int_0^{2\pi} f(r_\alpha)\,d\alpha + \frac{1}{4\pi}\int_0^{2\pi} f(sr_\alpha)\,d\alpha.$$

In particular, the projections p_i of 2.6 are:

$$p_i x = \frac{n_i}{4\pi}\int_0^{2\pi} \chi_i(r_\alpha)^*\,\rho_{r_\alpha}(x)\,d\alpha + \frac{n_i}{4\pi}\int_0^{2\pi} \chi_i(sr_\alpha)^*\,\rho_{sr_\alpha}(x)\,d\alpha.$$

Realizations of D_α *as a group of rigid motions in 3-space*

There are two of these:

(a) The usual realization (denoted D_∞ in Eyring [5]). Rotations are taken around Oz and reflections with respect to lines of the plane Oxy passing through O.

(b) The realization by means of the group $C_{\infty v}$ (notations of Eyring [5]): the reflections are taken with respect to planes passing through Oz, instead of lines of Oxy.

Irreducible representations of the group D_∞

They are constructed like those for D_n. There are first two representations of degree 1, with characters ψ_1 and ψ_2 given by the table:

	r_α	sr_α
ψ_1	1	1
ψ_2.	1	-1

There is a series of irreducible representations ρ^h of degree 2 ($h = 1, 2, \ldots$) defined by the formulas:

$$\rho^h(r_\alpha) = \begin{pmatrix} e^{ih\alpha} & 0 \\ 0 & e^{-ih\alpha} \end{pmatrix}, \qquad \rho^h(sr_\alpha) = \begin{pmatrix} 0 & e^{-ih\alpha} \\ e^{ih\alpha} & 0 \end{pmatrix}.$$

Their characters $\chi_1, \chi_2, \ldots$ have the following values:

$$\chi_h(r_\alpha) = 2\cos(h\alpha), \qquad \chi_h(sr_\alpha) = 0.$$

It can be shown that these are *all the irreducible representations* of D_∞ (up to isomorphism).

5.6 The group $D_{\infty h}$

This group is the product $D_\infty \times I$; it can be realized as the group generated by D_∞ and the reflection ι through the origin. Its elements can be written uniquely in one of the four forms:

$$r_\alpha, \qquad sr_\alpha, \qquad \iota r_\alpha, \qquad \iota sr_\alpha.$$

As a topological space, it is the union of four disjoint circles. The *invariant measure* of $D_{\infty h}$ is $(1/8\pi)\,d\alpha$. As above, this means that the average $\int_G f(t)\,dt$ of a function f on $D_{\infty h}$ is given by:

$$\int_G f(t)\,dt = \frac{1}{8\pi}\int_0^{2\pi} f(r_\alpha)\,d\alpha + \frac{1}{8\pi}\int_0^{2\pi} f(sr_\alpha)\,d\alpha + \frac{1}{8\pi}\int_0^{2\pi} f(\iota r_\alpha)\,d\alpha$$
$$+ \frac{1}{8\pi}\int_0^{2\pi} f(\iota sr_\alpha)\,d\alpha.$$

We leave it to the reader to derive the explicit expressions for the projections p_i of 2.6.

As in the case of D_{nh}, the irreducible representations of $D_{\infty h}$ come in pairs from D_∞: each character χ of D_∞ gives rise to two characters χ_g and χ_u of $D_{\infty h}$.

So, for example, the character χ_3 of D_∞ gives:

	r_α	sr_α	ιr_α	ιsr_α
χ_{3g}	$2\cos 3\alpha$	0	$2\cos 3\alpha$	0
χ_{3u}	$2\cos 3\alpha$	0	$-2\cos 3\alpha$	0

5.7 The alternating group $\mathfrak{A}_4$

This is the group of even permutations of a set $\{a,b,c,d\}$ having 4 elements; it is isomorphic to the group of rotations in $\mathbf{R}^3$ which stabilize a regular tetrahedron with barycenter the origin. It has 12 elements:

the identity element 1;
3 elements of order 2, $x = (ab)(cd)$, $y = (ac)(bd)$, $z = (ad)(bc)$, which correspond to reflections of the tetrahedron through lines joining the midpoints of two opposite edges;
8 elements of order 3: (abc), (acb), ..., (bcd), which correspond to rotations of $\pm 120°$ with respect to lines joining a vertex to the barycenter of the opposite face.

We denote by (abc) the cyclic permutation $a \mapsto b$, $b \mapsto c$, $c \mapsto a$, $d \mapsto d$; likewise, $(ab)(cd)$ denotes the permutation $a \mapsto b$, $b \mapsto a$, $c \mapsto d$, $d \mapsto c$, product of the transpositions (ab) and (cd).

Set $t = (abc)$, $K = \{1, t, t^2\}$ and $H = \{1, x, y, z\}$. We have

$$txt^{-1} = z, \qquad tzt^{-1} = y, \qquad tyt^{-1} = x;$$

moreover H and K are subgroups of $\mathfrak{A}_4$, H is normal, and $H \cap K = \{1\}$. It is easy to see that each element of $\mathfrak{A}_4$ *can be written uniquely as a product* $h \cdot k$, with $h \in H$ and $k \in K$.

One also says that $\mathfrak{A}_4$ is the *semidirect product* of K by the normal subgroup H; note that this is not a direct product, because the elements of K do not commute with those of H.

There are 4 *conjugacy classes in* $\mathfrak{A}_4$: $\{1\}$, $\{x,y,z\}$, $\{t, tx, ty, tz\}$, and $\{t^2, t^2x, t^2y, t^2z\}$, hence 4 irreducible characters. There are three characters of degree 1, corresponding to the three characters χ_0, χ_1, and χ_2 of the group K (cf. 5.1) extended to $\mathfrak{A}_4$ by setting $\chi_i(h \cdot k) = \chi_i(k)$ for $h \in H$ and $k \in K$. The last character ψ is determined, for example, by means of cor. 2

to prop. 5; it is found to be the character of the natural representation of $\mathfrak{A}_4$ in $\mathbf{R}^3$ (extended to $\mathbf{C}^3$ by linearity). Thus we have the following character table for $\mathfrak{A}_4$:

	1	x	t	t^2
χ_0	1	1	1	1
χ_1	1	1	w	w^2
χ_2	1	1	w^2	w
ψ	3	-1	0	0

with

$$w = e^{2\pi i/3} = -\frac{1}{2} + i\frac{\sqrt{3}}{2}.$$

EXERCISE

5.4. Set $\theta(1) = \theta(x) = 1$ and $\theta(y) = \theta(z) = -1$; this is a representation of degree 1 of H. The representation of $\mathfrak{A}_4$ induced by θ (cf. 3.3) is of degree 3; show that it is irreducible and has character ψ.

5.8 The symmetric group $\mathfrak{S}_4$

This is the group of all permutations of $\{a, b, c, d\}$; it is isomorphic to the group of all rigid motions which stabilize a regular tetrahedron. It has 24 elements, partitioned into 5 conjugacy classes:

the identity element 1;
6 transpositions: (ab), (ac), (ad), (bc), (bd), (cd);
the 3 elements of order 2 in $\mathfrak{A}_4$: $x = (ab)(cd), y = (ac)(bd), z = (ad)(bc)$;
8 elements of order 3: $(abc), \ldots, (bcd)$;
6 elements of order 4: $(abcd)$, $(abdc)$, $(acbd)$, $(acdb)$, $(adbc)$, $(adcb)$.

Let $\mathrm{H} = \{1, x, y, z\}$ and let L be the group of permutations which leave d fixed. We see, as in the preceding section, that $\mathfrak{S}_4$ is the *semidirect product* of L by the normal subgroup H. Each representation ρ of L is extended to a representation of $\mathfrak{S}_4$ by the formula $\rho(h \cdot l) = \rho(l)$ for $h \in \mathrm{H}$, $l \in \mathrm{L}$. This gives three irreducible representations of $\mathfrak{S}_4$ (cf. 2.5), of degrees 1, 1, and 2. On the other hand, the natural representation of $\mathfrak{S}_4$ in $\mathbf{C}^3$ is irreducible (since its restriction to $\mathfrak{A}_4$ is), and the same is true of its tensor product by the non-trivial representation of degree 1 of $\mathfrak{S}_4$. Whence the following character table for $\mathfrak{S}_4$:

	1	(ab)	$(ab)(cd)$	(abc)	$(abcd)$
χ_0	1	1	1	1	1
ε	1	-1	1	1	-1
θ	2	0	2	-1	0
ψ	3	1	-1	0	-1
$\varepsilon\psi$	3	-1	-1	0	1

Note that the values of the characters of $\mathfrak{S}_4$ are *integers*; this is a general property of representations of symmetric groups (cf. 13.1).

5.9 The group of the cube

Consider in $\mathbf{R}^3$ the cube C whose vertices are the points (x, y, z) with $x = \pm 1, y = \pm 1$, and $z = \pm 1$. Let G be the group of isomorphisms of $\mathbf{R}^3$ onto itself which stabilize the cube C, i.e., which permute its eight vertices. This group G can be described in several ways:

(i) The group G contains the group $\mathfrak{S}_3$ of permutations of $\{x, y, z\}$ as well as the group M of order 8 consisting of the transformations

$$(x, y, z) \mapsto (\pm x, \pm y, \pm z).$$

One checks easily that G is the *semidirect product* of $\mathfrak{S}_3$ by the normal subgroup M; its order is $6 \cdot 8 = 48$.

(ii) Denote by ι the reflection $(x, y, z) \mapsto (-x, -y, -z)$ through the origin. Let T be the tetrahedron whose vertices are the points $(1, 1, 1)$, $(1, -1, -1)$, $(-1, 1, -1)$, $(-1, -1, 1)$, and let $T' = \iota T$; each vertex of C is a vertex of T or of T′. Let S(T) be the group of isomorphisms of $\mathbf{R}^3$ onto itself which stabilize T; for $s \in S(T)$ we have $sT' = s\iota T = \iota s T = T'$, which shows that s stabilizes the set of vertices of C, and thus belongs to G. Consequently $S(T) \subset G$, and we see immediately that G is the *direct product of* S(T) *with the group* $I = \{1, \iota\}$. Since $S(T) = \mathfrak{S}_4$, the irreducible characters of G are obtained from those of $\mathfrak{S}_4$ in *pairs*, just as those of D_{nh} are obtained from those of D_n. Thus there are 4 irreducible characters of degree 1, 2 of degree 2, and 4 of degree 3; their exact description is left to the reader.

Exercises

5.4. Recover the semidirect decomposition $G = \mathfrak{S}_3 \cdot M$ from the decompositions $G = \mathfrak{S}_4 \times I$ and $\mathfrak{S}_4 = \mathfrak{S}_3 \cdot H$ (cf. 5.8).

5.5 Let G_+ be the subgroup of G consisting of elements with determinant 1 (the group of *rotations* of the cube). Show that, if G is decomposed into $S(T) \times I$, the projection $G \to S(T)$ defines an *isomorphism* of G_+ *onto* $S(T) = \mathfrak{S}_4$.

Bibliography: Part I

The representation theory of finite groups is discussed in a number of books. We mention first a classic:

[1] H. Weyl. *The Theory of Groups and Quantum Mechanics*. Dover Publ., 1931.

See also:

[2] M. Hall. *The Theory of Groups*. Macmillan, New York, 1959.

[3] G. G. Hall. *Applied Group Theory*. Mathematical Physics Series, Longmans, 1967.

A discussion of induced representations, together with their applications, is found in:

[4] A. J. Coleman. Induced and subduced representations. *Group Theory and Its Applications*, edited by M. Loebl. Academic Press, New York, 1968.

For the groups of rigid motions in $\mathbf{R}^3$, standard notations and tables of characters, see:

[5] H. Eyring, J. Walter, and G. Kimball. *Quantum Chemistry*. John Wiley and Sons, New York, 1944.

(The tables are in Appendix VII, pp. 376–388.)

For compact groups, see [1], [4], as well as:

[6] L. Loomis. *An Introduction to Abstract Harmonic Analysis*. Van Nostrand, New York, 1953.

The reader who is interested in the history of character theory may consult the original papers of Frobenius and Schur:

[7] F. G. Frobenius. *Gesammelte Abhandlungen*, Bd. III. Springer-Verlag, 1969.

[7′] I. Schur. *Gesammelte Abhandlungen*, Bd.I. Springer-Verlag, 1973.

[7″] T. Hawkins. *New Light on Frobenius' Creation of the Theory of Group Characters*, Archive for History of Exact Sciences, 12 (1974), pp. 217–243.

II
REPRESENTATIONS IN CHARACTERISTIC ZERO

Unless explicitly stated otherwise, all groups are assumed to be finite, and all vector spaces (resp., all modules) are assumed to be of finite dimension (resp., finitely generated).

In Ch. 6 to 11 (except for 6.1) the ground field is the field **C** of complex numbers.

CHAPTER 6

The group algebra

6.1 Representations and modules

Let G be a group of finite order g, and let K be a commutative ring. We denote by K[G] the *algebra of* G *over* K; this algebra has a basis indexed by the elements of G, and most of the time we identify this basis with G. Each element f of K[G] can then be uniquely written in the form

$$f = \sum_{s \in G} a_s s, \quad \text{with } a_s \in K,$$

and multiplication in K[G] extends that in G.

Let V be a K-module and let ρ: G $\to$ **GL**(V) be a linear representation of G in V. For $s \in G$ and $x \in V$, set $sx = \rho_s x$; by linearity this defines fx, for $f \in K[G]$ and $x \in V$. Thus V is endowed with the structure of a *left* K[G]-*module*; conversely, such a structure defines a linear representation of G in V. In what follows we will indiscriminately use the terminology "linear representation" or "module."

Proposition 9. *If* K *is a field of characteristic zero, the algebra* K[G] *is semisimple.*

(For the basic facts on semisimple algebras, see, for example, Bourbaki [8] or Lang [10].)

To say that K[G] is a semisimple algebra is equivalent to saying that each K[G]-module V is semisimple, i.e., that each submodule V′ of V is a direct factor in V as a K[G]-module. This is proved by the same argument of *averaging* as that in 1.3: we choose first a K-linear projection p of V onto V′, then form the average $p^0 = (1/g) \sum_{s \in G} sps^{-1}$ of its transforms by G.

The projection p^0 thus obtained is K[G]-linear, which implies that V′ is a direct factor of V as a K[G]-module. □

Corollary. *The algebra* K[G] *is a product of matrix algebras over skew fields of finite degree over* K.

This is a consequence of the structure theorem for semisimple algebras (*loc. cit.*).

EXERCISE

6.1. Let K be a field of characteristic $p > 0$. Show that the following two properties are equivalent:
(i) K[G] is semisimple.
(ii) p does not divide the order g of G.
(The fact that (ii) ⇒ (i) is proved as above. To prove the converse, show that, if p divides g, the ideal of K[G] consisting of the $\sum a_s s$ with $\sum a_s = 0$ is not a direct factor (as a module) of K[G].)

6.2 Decomposition of C[G]

Henceforth we take $\mathrm{K} = \mathbf{C}$ (though any algebraically closed field of characteristic zero would do as well), so that each skew field of finite degree over **C** is equal to **C**. The corollary to prop. 9 then shows that **C**[G] *is a product of matrix algebras* $\mathbf{M}_{n_i}(\mathbf{C})$. More precisely, let $\rho_i\colon \mathrm{G} \to \mathbf{GL}(\mathrm{W}_i)$, $1 \leqslant i \leqslant h$, be the distinct irreducible representations of G (up to isomorphism), and set $n_i = \dim(\mathrm{W}_i)$, so that the ring $\mathrm{End}(\mathrm{W}_i)$ of endomorphisms of W_i is isomorphic to $\mathbf{M}_{n_i}(\mathbf{C})$. The map $\rho_i\colon \mathrm{G} \to \mathbf{GL}(\mathrm{W}_i)$ extends by linearity to an algebra homomorphism $\tilde{\rho}_i\colon \mathbf{C}[\mathrm{G}] \to \mathrm{End}(\mathrm{W}_i)$; the family $(\tilde{\rho}_i)$ defines a homomorphism

$$\tilde{\rho}\colon \mathbf{C}[\mathrm{G}] \to \prod_{i=1}^{i=h} \mathrm{End}(\mathrm{W}_i) \simeq \prod_{i=1}^{i=h} \mathbf{M}_{n_i}(\mathbf{C}).$$

Proposition 10. *The homomorphism* $\tilde{\rho}$ *defined above is an isomorphism.*

This is a general property of semisimple algebras. In the present case, it can be verified in the following way: First, $\tilde{\rho}$ is *surjective*. Otherwise there would exist a nonzero linear form on $\prod \mathbf{M}_{n_i}(\mathbf{C})$ vanishing on the image of $\tilde{\rho}$; this would give a nontrivial relation on the coefficients of the representations ρ_i, which is impossible because of the orthogonality formulas of 2.2. On the other hand, **C**[G] and $\prod \mathbf{M}_{n_i}(\mathbf{C})$ both have dimension $g = \sum n_i^2$, cf. 2.4; so since $\tilde{\rho}$ is surjective, it must be bijective. □

It is possible to describe the isomorphism which is the inverse of $\tilde{\rho}$:

Proposition 11 (Fourier inversion formula). *Let $(u_i)_{1 \leqslant i \leqslant h}$ be an element of $\prod \mathrm{End}(W_i)$, and let $u = \sum_{s \in G} u(s)s$ be the element of* C[G] *such that $\tilde{\rho}_i(u) = u_i$ for all i. The* sth *coefficient $u(s)$ of u is given by the formula*

$$u(s) = \frac{1}{g} \sum_{i=1}^{i=h} n_i \mathrm{Tr}_{W_i}(\rho_i(s^{-1})u_i), \quad \textit{where } n_i = \dim(W_i).$$

By linearity it is enough to check the formula when u is equal to an element t of G. We have then

$$u(s) = \delta_{st} \qquad \text{and} \quad \mathrm{Tr}_{W_i}(\rho_i(s^{-1})u_i) = \chi_i(s^{-1}t),$$

where χ_i is the irreducible character of G corresponding to W_i. Thus it remains to show that

$$\delta_{st} = \frac{1}{g} \sum_{i=1}^{i=h} n_i \chi_i(s^{-1}t),$$

which is a consequence of cor. 1 and 2 of prop. 5 of 2.4. □

EXERCISES

6.2. (*Plancherel formula.*) Let $u = \sum u(s)s$ and $v = \sum v(s)s$ be two elements of C[G], and put $\langle u, v \rangle = g \sum_{s \in G} u(s^{-1})v(s)$. Prove the formula

$$\langle u, v \rangle = \sum_{i=1}^{i=h} n_i \mathrm{Tr}_{W_i}(\tilde{\rho}_i(uv)).$$

[Reduce to the case where u and v belong to G.]

6.3. Let U be a finite subgroup of the multiplicative group of C[G] which contains G. Let $u = \sum u(s)s$ and $u' = \sum u'(s)s$ be two elements of U such that $u \cdot u' = 1$; let u_i (resp. u'_i) be the image of u (resp. u') in $\mathrm{End}(W_i)$ under $\tilde{\rho}_i$.

(a) Show that the eigenvalues of $\rho_i(s^{-1})u_i = \tilde{\rho}_i(s^{-1}u)$ are roots of unity. Conclude that, for all $s \in$ G and all i, we have

$$\mathrm{Tr}_{W_i}(\rho_i(s^{-1})u_i)^* = \mathrm{Tr}_{W_i}(u'_i \rho_i(s)) = \mathrm{Tr}_{W_i}(\rho_i(s)u'_i),$$

whence, applying prop. 11, $u(s)^* = u'(s^{-1})$.

(b) Show that $\sum_{s \in G} |u(s)|^2 = 1$ [use (*a*)].

(c) Suppose that U is contained in Z[G] so that the $u(s)$ are integers. Show that the $u(s)$ are all zero except for one which is equal to ± 1. Conclude that U is contained in the group $\pm$G of elements of the form $\pm t$, with $t \in$ G.

(d) Suppose G is abelian. Show that each element of finite order in the multiplicative group of Z[G] is contained in $\pm$G (*Higman's theorem*).

6.3 The center of C[G]

This is the set of elements of C[G] which *commute* with all the elements in C[G] (or, what amounts to the same thing, with all the elements of G).

For c a conjugacy class of G, set $e_c = \sum_{s \in c} s$. One checks immediately that the e_c form a basis for the center of C[G]; the latter therefore has dimension h, where h is the *number of classes of* G, cf. 2.5. Let

$$\rho_i: G \to \mathbf{GL}(W_i)$$

be an irreducible representation of G with character χ_i and degree n_i, and let $\tilde{\rho}_i$: C[G] → End(W_i) be the corresponding algebra homomorphism (cf. 6.2).

Proposition 12. *The homomorphism $\tilde{\rho}_i$ maps the center of* C[G] *into the set of homotheties of* W_i *and defines an algebra homomorphism*

$$\omega_i: \text{Cent. } \mathbf{C}[G] \to \mathbf{C}.$$

If $u = \sum u(s)s$ *is an element of* Cent. C[G], *we have*

$$\omega_i(u) = \frac{1}{n_i} \mathrm{Tr}_{W_i}(\tilde{\rho}_i(u)) = \frac{1}{n_i} \sum_{s \in G} u(s)\chi_i(s).$$

This is just a reformulation of prop. 6 of 2.5.

Proposition 13. *The family* $(\omega_i)_{1 \leqslant i \leqslant h}$ *defines an isomorphism of* Cent. C[G] *onto the algebra* $\mathbf{C}^h = \mathbf{C} \times \cdots \times \mathbf{C}$.

If we identify C[G] with the product of the End(W_i), the center of C[G] becomes the product of the centers of the End(W_i). But the center of End(W_i) consists of homotheties. We thus get an isomorphism of Cent. C[G] onto $\mathbf{C} \times \cdots \times \mathbf{C}$, and it is immediate that it is the one of prop. 13. □

EXERCISES

6.4. Set

$$p_i = \frac{n_i}{g} \sum_{s \in G} \chi_i(s^{-1})s.$$

Show that the $p_i (1 \leqslant i \leqslant h)$ form a basis of Cent. C[G] and that $p_i^2 = p_i$, $p_i p_j = 0$ for $i \neq j$, and $p_1 + \cdots + p_h = 1$. Hence obtain another proof of th. 8 of 2.6. Show that $\omega_i(p_j) = \delta_{ij}$.

6.5. Show that each homomorphism of Cent. C[G] into C is equal to one of the ω_i.

6.4 Basic properties of integers

Let R be a commutative ring and let $x \in$ R. We say that x is *integral over* **Z** if there exists an integer $n \geqslant 1$ and elements $a_1, \ldots, a_n$ of **Z** such that

$$x^n + a_1 x^{n-1} + \cdots + a_n = 0.$$

A complex number which is integral over $\mathbf{Z}$ is called an *algebraic integer.* Each root of unity is an algebraic integer. *If $x \in \mathbf{Q}$ is an algebraic integer, we have* $x \in \mathbf{Z}$; otherwise we could write x in the form p/q, with $p, q \in \mathbf{Z}$, $q \geqslant 2$ and p, q relatively prime. The equation $(*)$ would then give

$$p^n + a_1 qp^{n-1} + \cdots + a_n q^n = 0,$$

hence $p^n \equiv 0$ (mod. q) contradicting the fact that p and q are relatively prime.

Proposition 14. *Let x be an element of a commutative ring* R. *The following properties are equivalent:*

(i) *x is integral over* $\mathbf{Z}$.
(ii) *The subring* $\mathbf{Z}[x]$ *of* R *generated by x is finitely generated as a* $\mathbf{Z}$-*module.*
(iii) *There exists a finitely generated sub-*$\mathbf{Z}$*-module of* R *which contains* $\mathbf{Z}[x]$.

The equivalence of (ii) and (iii) follows from the fact that a submodule of a finitely generated $\mathbf{Z}$-module is finitely generated, since $\mathbf{Z}$ is *noetherian.* On the other hand, if x satisfies an equation

$$x^n + a_1 x^{n-1} + \cdots + a_n = 0, \quad \text{with } a_i \in \mathbf{Z},$$

the sub-$\mathbf{Z}$-module of R generated by $1, x, \ldots, x^{n-1}$ is stable under multiplication by x, and thus coincides with $\mathbf{Z}[x]$, which proves (i) $\Rightarrow$ (ii). Conversely, suppose (ii) is satisfied, and denote by R_n the sub-$\mathbf{Z}$-module of R generated by $1, x, \ldots, x^{n-1}$. The R_n form an increasing sequence, and their union is $\mathbf{Z}[x]$; since $\mathbf{Z}[x]$ is finitely generated we must have $\mathrm{R}_n = \mathbf{Z}[x]$ for n sufficiently large. This shows that x^n is a linear combination with integer coefficients of $1, x, \ldots, x^{n-1}$, whence (i). □

Corollary 1. *If* R *is a finitely generated* $\mathbf{Z}$*-module, each element of* R *is integral over* $\mathbf{Z}$.

This follows from the implication (iii) $\Rightarrow$ (i). □

Corollary 2. *The elements of* R *which are integral over* $\mathbf{Z}$ *form a subring of* R.

Let $x, y \in \mathrm{R}$; if x, y are integral over $\mathbf{Z}$, the rings $\mathbf{Z}[x]$ and $\mathbf{Z}[y]$ are finitely generated over $\mathbf{Z}$. The same is then true of their tensor product $\mathbf{Z}[x] \otimes \mathbf{Z}[y]$ and of its image $\mathbf{Z}[x,y]$ in R. Thus all the elements of $\mathbf{Z}[x,y]$ are integral over $\mathbf{Z}$. □

Remark. In the preceding definitions and results it is possible to replace $\mathbf{Z}$ by an arbitrary commutative noetherian ring; for (i) $\Leftrightarrow$ (ii) it is not even necessary to assume the ring is noetherian.

6.5 Integrality properties of characters. Applications

Proposition 15. *Let χ be the character of a representation ρ of a finite group G. Then $\chi(s)$ is an algebraic integer for each $s \in$ G.*

Indeed $\chi(s)$ is the trace of $\rho(s)$, hence is the sum of eigenvalues of $\rho(s)$, which are roots of unity.

Proposition 16. *Let $u = \sum u(s)s$ be an element of* Cent. $\mathbf{C}[G]$ *such that the $u(s)$ are algebraic integers. Then u is integral over* $\mathbf{Z}$.

(This statement makes sense because Cent. $\mathbf{C}[G]$ is a commutative ring.)

Let $c_i (1 \leqslant i \leqslant h)$ be the conjugacy classes of G and put $e_i = \sum_{s \in c_i} s$, cf. 6.3. For $s_i \in c_i$ we can write u in the form $u = \sum_{i=1}^{i=h} u(s_i)e_i$. In view of cor. 2 to prop. 14, it suffices to show that the e_i *are integral over* $\mathbf{Z}$. But this is clear since each product $e_i e_j$ is a linear combination with integer coefficients of the e_k. The subgroup $R = \mathbf{Z}e_1 \oplus \cdots \oplus \mathbf{Z}e_h$ of Cent. $\mathbf{C}[G]$ is thus a subring; as it is finitely generated over $\mathbf{Z}$, each of its elements is integral over $\mathbf{Z}$ (cor. 1 to prop. 14). The result follows. □

Corollary 1. *Let ρ be an irreducible representation of G of degree n and character χ. If u is as above, then the number $(1/n) \sum_{s \in G} u(s)\chi(s)$ is an algebraic integer.*

Indeed, this number is the image of u under the homomorphism

$$\omega\colon \text{Cent}.\, \mathbf{C}[G] \to \mathbf{C}$$

associated with ρ (cf. prop. 12). As u is integral over $\mathbf{Z}$, the same is true of its image under ω.

Corollary 2. *The degrees of the irreducible representations of* G *divide the order of* G.

Let g be the order of G. We apply cor. 1 to the element $u = \sum_{s \in G} \chi(s^{-1})s$, which is legitimate since χ is a class function and since the $\chi(s)$ are algebraic integers (prop. 15); we obtain that the number

$$\frac{1}{n} \sum_{s \in G} \chi(s^{-1})\chi(s) = \frac{g}{n}\langle \chi, \chi \rangle = \frac{g}{n}$$

is an algebraic integer. Since this number is rational, it follows that it belongs to $\mathbf{Z}$, i.e., that n divides g. □

Corollary 2 can be strengthened somewhat (cf. 8.1, cor. to prop. 24). Here is a first result in this direction:

Proposition 17. *Let* C *be the center of* G. *The degrees of the irreducible representations of* G *divide* (G: C).

Let g be the order of G and c that of C, and let ρ: G $\rightarrow$ **GL**(W) be an irreducible representation of G of degree n. If $s \in$ C, $\rho(s)$ commutes with all the $\rho(t)$, $t \in$ G; so by Schur's lemma, $\rho(s)$ is a homothety. If we denote it by $\lambda(s)$, the map λ: $s \mapsto \lambda(s)$ is a homomorphism of C into $\mathbf{C}^*$. Let m be an integer $\geqslant 0$, and form the tensor product

$$\rho^m: \mathrm{G}^m \rightarrow \mathbf{GL}(\mathrm{W} \otimes \cdots \otimes \mathrm{W})$$

of m copies of the representation ρ; this is an irreducible representation of the group $\mathrm{G}^m = \mathrm{G} \times \cdots \times \mathrm{G}$, cf. 3.2 th. 10. The image under ρ^m of an element $(s_1, \ldots, s_m)$ of C^m is the homothety of ratio $\lambda(s_1 \cdots s_m)$. The subgroup H of C^m consisting of the $(s_1, \ldots, s_m)$ such that $s_1 \cdots s_m = 1$ acts trivially on $\mathrm{W} \otimes \cdots \otimes \mathrm{W}$, so that by passing to the quotient we obtain an irreducible representation of G^m/H. In view of cor. 2 to prop. 16, it follows that the degree n^m of this representation divides the order g^m/c^{m-1} of G^m/H. We have then $(g/cn)^m \in c^{-1}\mathbf{Z}$ for all m, which implies that (g/cn) is an integer (cf. prop. 14, for example).

(This proof is due to J. Tate.) □

EXERCISES

6.6. Show that the ring $\mathbf{Z}e_1 \oplus \cdots \oplus \mathbf{Z}e_h$ is the center of $\mathbf{Z}[\mathrm{G}]$.

6.7. Let ρ be an irreducible representation of G of degree n and with character χ. If $s \in$ G, show that $|\chi(s)| \leqslant n$, and that equality holds if and only if $\rho(s)$ is a homothety [observe that $\chi(s)$ is a sum of n roots of unity]. Conclude that $\rho(s) = 1 \Leftrightarrow \chi(s) = n$.

6.8. Let $\lambda_1, \ldots, \lambda_n$ be roots of unity, and let $a = \frac{1}{n}\sum \lambda_i$. Show that, if a is an algebraic integer, we have either $a = 0$, or $\lambda_1 = \cdots = \lambda_n = a$. [Let A be the product of the conjugates of a over **Q**; show that $|\mathrm{A}| \leqslant 1$.]

6.9. Let ρ be an irreducible representation of G of degree n and with character χ. Let $s \in$ G and $c(s)$ be the number of elements in the conjugacy class of s. Show that $(c(s)/n)\chi(s)$ is an algebraic integer [apply cor. 1 to prop. 16, taking for u the sum of the conjugates of s]. Show that if $c(s)$ and n are relatively prime and if $\chi(s) \neq 0$, then $\rho(s)$ is a homothety [Observe that $(1/n)\chi(s)$ is an algebraic integer, and apply ex. 6.8].

6.10. Let $s \in$ G, $s \neq 1$. Suppose that the number of elements $c(s)$ of the conjugacy class containing s is a power of a prime number p. Show that there exists an irreducible character χ, not equal to the unit character, such that $\chi(s) \neq 0$ and $\chi(1) \not\equiv 0$ (mod.p). [Use the formula $1 + \sum_{\chi \neq 1} \chi(1)\chi(s) = 0$, cf. cor. 2 to prop. 5 to show that the number $1/p$ would be an algebraic integer if no such character χ existed.] Let ρ be a representation with character χ, and show that $\rho(s)$ is a homothety [use ex. 6.9]. Conclude that, if N is the kernel of ρ, we have N $\neq$ G, and the image of s in G/N belongs to the center of G/N.

CHAPTER 7

Induced representations; Mackey's criterion

7.1 Induction

Let H be a subgroup of a group G and R a system of left coset representatives for H. Let V be a **C**[G]-module and let W be a sub-**C**[H]-module of V. Recall (cf. 3.3) that the module V (or the representation V) is said to be *induced* by W if we have $V = \bigoplus_{s \in R} sW$, i.e., if V is a direct sum of the images sW, $s \in R$ (a condition which is independent of the choice of R). This property can be reformulated in the following way:

Let

$$W' = \mathbf{C}[G] \otimes_{\mathbf{C}[H]} W$$

be the **C**[G]-module obtained from W by scalar extension from **C**[H] to **C**[G]. The injection $W \to V$ extends by linearity to a **C**[G]-homomorphism $i: W' \to V$.

Proposition 18. *In order that* V *be induced by* W, *it is necessary and sufficient that the homomorphism*

$$i: \mathbf{C}[G] \otimes_{\mathbf{C}[H]} W \to V$$

be an isomorphism.

This is a consequence of the fact that the elements of R form a basis of **C**[G] considered as a right **C**[H]-module.

Remarks

(1) This characterization of the representation induced by W makes it obvious that the induced representation *exists* and is *unique* (cf. 3.3, th. 11).

In what follows, the representation of G induced by W will be denoted by $\mathrm{Ind}_H^G(W)$, or simply $\mathrm{Ind}(W)$ if there is no danger of confusion.

(2) If V is induced by W and if E is a $\mathbf{C}[G]$-module, we have a canonical isomorphism

$$\mathrm{Hom}^H(W, E) \cong \mathrm{Hom}^G(V, E),$$

where $\mathrm{Hom}^G(V, E)$ denotes the vector space of $\mathbf{C}[G]$-homomorphisms of V into E, and $\mathrm{Hom}^H(W, E)$ is defined similarly. This follows from an elementary property of tensor products (see also 3.3, lemma 1).

(3) Induction is *transitive*: if G is a subgroup of a group K, we have

$$\mathrm{Ind}_G^K(\mathrm{Ind}_H^G(W)) \cong \mathrm{Ind}_H^K(W).$$

This can be seen directly, or by using the associativity of the tensor product.

Proposition 19. *Let* V *be a* $\mathbf{C}[G]$*-module which is a direct sum* $V = \bigoplus_{i \in I} W_i$ *of vector subspaces permuted transitively by* G. *Let* $i_0 \in I$, $W = W_{i_0}$ *and let* H *be the stabilizer of* W *in* G (*i.e., the set of all* $s \in G$ *such that* $sW = W$). *Then* W *is stable under the subgroup* H *and the* $\mathbf{C}[G]$*-module* V *is induced by the* $\mathbf{C}[H]$*-module* W.

This is clear.

Remark. In order to apply proposition 19 to an irreducible representation $V = \oplus W_i$ of G, it is enough to check that the W_i are permuted among themselves by G; the transitivity condition is automatic, because each orbit of G in the set of W_i's defines a subrepresentation of V.

EXAMPLE. When the W_i are of dimension 1, the representation V is said to be *monomial.*

7.2 The character of an induced representation; the reciprocity formula

We keep the preceding notation. If f is a class function on H, consider the function f' on G defined by the formula

$$f'(s) = \frac{1}{h} \sum_{\substack{t \in G \\ t^{-1}st \in H}} f(t^{-1}st) \qquad \text{where } h = \mathrm{Card}(H).$$

We say that f' is *induced* by f and denote it by either $\mathrm{Ind}_H^G(f)$ or $\mathrm{Ind}(f)$.

Proposition 20.

(i) *The function* Ind(f) *is a class function on* G.
(ii) *If f is the character of a representation* W *of* H, Ind(f) *is the character of the induced representation* Ind(W) *of* G.

Assertion (ii) has already been proved (3.3, th. 12). Assertion (i) is proved by a direct calculation or can be obtained from (ii) and the observation that each class function is a linear combination of characters. □

Recall that, for φ_1 and φ_2 two class functions on G, we set

$$\langle \varphi_1, \varphi_2 \rangle = \frac{1}{g} \sum_{s \in G} \varphi_1(s^{-1})\varphi_2(s), \quad \text{where } g = \text{Card}(G),$$

cf. 2.2; when we wish to be more explicit about the group G, we write $\langle \varphi_1, \varphi_2 \rangle_G$ instead of $\langle \varphi_1, \varphi_2 \rangle$.

Also, if V_1 and V_2 are two **C**[G]-modules, we set

$$\langle V_1, V_2 \rangle_G = \dim . \text{Hom}^G(V_1, V_2).$$

Lemma 2. *If φ_1 and φ_2 are the characters of V_1 and V_2, we have*

$$\langle \varphi_1, \varphi_2 \rangle_G = \langle V_1, V_2 \rangle_G.$$

Decomposing V_1 and V_2 into direct sums, we can assume that they are irreducible, in which case the lemma follows from the orthogonality formulas for characters (2.3, th. 3). □

If φ (resp. V) is a function on G (resp. a representation of G), we denote by Res φ (resp. Res V) its *restriction* to the subgroup H.

Theorem 13 (Frobenius reciprocity). *If ψ is a class function on* H *and φ a class function on* G, *we have*

$$\langle \psi, \text{Res } \varphi \rangle_H = \langle \text{Ind } \psi, \varphi \rangle_G.$$

Since each class function is a linear combination of characters, we can assume that ψ is the character of a **C**[H]-module W and φ is the character of a **C**[G]-module E. In view of lemma 2, it is enough to show that

$$(*) \qquad \langle W, \text{Res } E \rangle_H = \langle \text{Ind } W, E \rangle_G,$$

that is,

$$\dim.\text{Hom}^H(W, \text{Res } E) = \dim.\text{Hom}^G(\text{Ind } W, E),$$

which follows from remark 2 in 7.1 (or from lemma 1 of 3.3, which amounts to the same thing). Of course it is also possible to prove theorem 13 by direct calculation. □

Remarks

(1) Theorem 13 expresses the fact that the maps Res and Ind are *adjoints* of each other.

(2) Instead of the bilinear form $\langle \alpha, \beta \rangle$, we can use the scalar product $(\alpha | \beta)$ defined in 2.3. We have the same formula:

$$(\psi | \text{Res } \varphi)_H = (\text{Ind } \psi | \varphi)_G.$$

(3) We mention also the following useful formula

$$\text{Ind}(\psi \cdot \text{Res } \varphi) = (\text{Ind } \psi) \cdot \varphi.$$

It can be checked by a simple calculation, or deduced from the formula $\text{Ind}(W) \otimes E \cong \text{Ind}(W \otimes \text{Res } E)$, cf. 3.3, example 5.

Proposition 21. *Let* W *be an irreducible representation of* H *and* E *an irreducible representation of* G. *Then the number of times that* W *occurs in* Res E *is equal to the number of times that E occurs in* Ind W

This follows from th. 13, applied to the character ψ of W and to the character φ of E (one may also apply formula (*)). □

Exercises

7.1. (Generalization of the concept of induced representation.) Let α: H → G be a homomorphism of groups (not necessarily injective), and let $\tilde{\alpha}$: **C**[H] → **C**[G] be the corresponding algebra homomorphism. If E is a **C**[G]-module we denote by $\text{Res}_\alpha E$ the **C**[H]-module obtained from E by means of $\tilde{\alpha}$; if φ is the character of E, that of $\text{Res}_\alpha E$ is $\text{Res}_\alpha \varphi = \varphi \circ \alpha$. If W is a **C**[H]-module, we denote by $\text{Ind}_\alpha W$ the **C**[G]-module $\mathbf{C}[G] \otimes_{\mathbf{C}[H]} W$, and if ψ is the character of W, we denote by $\text{Ind}_\alpha \psi$ the character of $\text{Ind}_\alpha W$.

(a) Show that we still have the reciprocity formula

$$\langle \psi, \text{Res}_\alpha \varphi \rangle_H = \langle \text{Ind}_\alpha \psi, \varphi \rangle_G.$$

(b) Assume that α is surjective and identify G with the quotient of H by the kernel N of α. Show that $\text{Ind}_\alpha W$ is isomorphic to the module obtained by having G = H/N act on the subspace of W consisting of the elements invariant under N. Deduce the formula

$$(\text{Ind}_\alpha \psi)(s) = \frac{1}{n} \sum_{\alpha(t)=s} \psi(t) \quad \text{where } n = \text{Card}(N).$$

7.2. Let H be a subgroup of G and let χ be the character of the permutation representation associated with G/H (cf. 1.2). Show that $\chi = \text{Ind}_H^G(1)$, and that $\psi = \chi - 1$ is the character of a representation of G; determine under what condition the latter representation is irreducible [use ex. 2.6, or apply the reciprocity formula].

7.3. Let H be a subgroup of G. Assume that for each $t \notin \mathrm{H}$ we have $\mathrm{H} \cap t\mathrm{H}t^{-1} = \{1\}$, in which case H is said to be a Frobenius subgroup of G. Denote by N the set of elements of G which are not conjugate to any element of H.

(a) Let $g = \mathrm{Card}(\mathrm{G})$ and let $h = \mathrm{Card}(\mathrm{H})$. Show that the number of elements of N is $(g/h) - 1$.

(b) Let f be a class function on H. Show that there exists a unique class function $\tilde{f}$ on G which extends f and takes the value $f(1)$ on N.

(c) Show that $\tilde{f} = \mathrm{Ind}_{\mathrm{H}}^{\mathrm{G}} f - f(1)\psi$, where ψ is the character $\mathrm{Ind}_{\mathrm{H}}^{\mathrm{G}}(1) - 1$ of G, cf. ex. 7.2.

(d) Show that $\langle f_1, f_2 \rangle_{\mathrm{H}} = \langle \tilde{f}_1, \tilde{f}_2 \rangle_{\mathrm{G}}$.

(e) Take f to be an irreducible character of H. Show, using (c) and (d), that $\langle \tilde{f}, \tilde{f} \rangle_{\mathrm{G}} = 1$, $\tilde{f}(1) \geq 0$, and that $\tilde{f}$ is a linear combination with integer coefficients of irreducible characters of G. Conclude that $\tilde{f}$ is an irreducible character of G. If ρ is a corresponding representation of G, show that $\rho(s) = 1$ for each $s \in \mathrm{N}$ [use ex. 6.7].

(f) Show that each linear representation of H extends to a linear representation of G whose kernel contains N. Conclude that $\mathrm{N} \cup \{1\}$ is a normal subgroup of G and that G is the semidirect product of H and $\mathrm{N} \cup \{1\}$ (*Frobenius' theorem*).

(g) Conversely, suppose G is the semidirect product of H and a normal subgroup A. Show that H is a Frobenius subgroup of G if and only if for each $s \in \mathrm{H} - \{1\}$ and each $t \in \mathrm{A} - \{1\}$, we have $sts^{-1} \neq t$ (i.e., H acts freely on $\mathrm{A} - \{1\}$). (If $\mathrm{H} \neq \{1\}$, this property implies that A is nilpotent, by a theorem of Thompson.)

7.3 Restriction to subgroups

Let H and K be two subgroups of G, and let $\rho\colon \mathrm{H} \to \mathbf{GL}(\mathrm{W})$ be a linear representation of H, and let $\mathrm{V} = \mathrm{Ind}_{\mathrm{H}}^{\mathrm{G}}(\mathrm{W})$ be the corresponding induced representation of G. We shall determine the restriction $\mathrm{Res}_{\mathrm{K}} \mathrm{V}$ of V to K.

First choose a set of representatives S for the (H, K) double cosets of G; this means that G is the disjoint union of the $\mathrm{K}s\mathrm{H}$ for $s \in \mathrm{S}$ (we could also write $s \in \mathrm{K}\backslash\mathrm{G}/\mathrm{H}$). For $s \in \mathrm{S}$, let $\mathrm{H}_s = s\mathrm{H}s^{-1} \cap \mathrm{K}$, which is a subgroup of K. If we set

$$\rho^s(x) = \rho(s^{-1}xs), \quad \text{for } x \in \mathrm{H}_s,$$

we obtain a homomorphism $\rho^s\colon \mathrm{H}_s \to \mathbf{GL}(\mathrm{W})$, and hence a linear representation of H_s, denoted W_s. Since H_s is a subgroup of K, the induced representation $\mathrm{Ind}_{\mathrm{H}_s}^{\mathrm{K}}(\mathrm{W}_s)$ is defined.

Proposition 22. *The representation* $\mathrm{Res}_{\mathrm{K}} \mathrm{Ind}_{\mathrm{H}}^{\mathrm{G}}(\mathrm{W})$ *is isomorphic to the direct sum of the representations* $\mathrm{Ind}_{\mathrm{H}_s}^{\mathrm{K}}(\mathrm{W}_s)$, *for* $s \in \mathrm{S} \simeq \mathrm{K}\backslash\mathrm{G}/\mathrm{H}$.

We know that V is the direct sum of the images xW, for $x \in G/H$. Let $s \in S$ and let $V(s)$ be the subspace of V generated by the images xW, for $x \in KsH$; the space V is a direct sum of the $V(s)$, and it is clear that $V(s)$ is stable under K. It remains to prove that $V(s)$ is K-isomorphic to $\mathrm{Ind}_{H_s}^{K}(W_s)$. But the subgroup of K consisting of the elements x such that $x(sW) = sW$ is evidently equal to H_s, and $V(s)$ is a direct sum of the images $x(sW)$, $x \in K/H_s$. Therefore $V(s) = \mathrm{Ind}_{H_s}^{K}(sW)$. Now it remains to check that sW is H_s-isomorphic to W_s, and this is immediate: the isomorphism is given by $s\colon W_s \to sW$. □

Remark. Since $V(s)$ depends only on the image of s in $K\backslash G/H$, we also see that the representation $\mathrm{Ind}_{H_s}^{K}(W_s)$ depends (up to isomorphism) only on the double coset of s.

7.4 Mackey's irreducibility criterion

We apply the preceding results to the case $K = H$. For $s \in G$, we still denote by H_s the subgroup $sHs^{-1} \cap H$ of H; the representation ρ of H defines a representation $\mathrm{Res}_s(\rho)$ by restriction to H_s, which should not be confused with the representation ρ^s defined in 7.3.

Proposition 23. *In order that the induced representation* $V = \mathrm{Ind}_H^G W$ *be irreducible, it is necessary and sufficient that the following two conditions be satisfied:*

(a) W *is irreducible.*
(b) *For each* $s \in G - H$ *the two representations* ρ^s *and* $\mathrm{Res}_s(\rho)$ *of* H_s *are disjoint.*

(Two representations V_1 and V_2 of a group K are said to be *disjoint* if they have no irreducible component in common, i.e., if $\langle V_1, V_2\rangle_K = 0$.)

In order that V be irreducible, it is necessary and sufficient that $\langle V, V\rangle_G = 1$. But, according to Frobenius reciprocity, we have:

$$\langle V, V\rangle_G = \langle W, \mathrm{Res}_H V\rangle_H .$$

However, from 7.3 we have:

$$\mathrm{Res}_H V = \bigoplus_{s\in H\backslash G/H} \mathrm{Ind}_{H_s}^{H}(\rho^s).$$

Once more applying the Frobenius formula, we obtain:

$$\langle V, V\rangle_G = \sum_{s\in H\backslash G/H} d_s, \quad \text{with } d_s = \langle \mathrm{Res}_s(\rho), \rho^s\rangle_{H_s}.$$

For $s = 1$ we have $d_s = \langle \rho, \rho \rangle \geqslant 1$. In order that $\langle V, V \rangle_G = 1$, it is thus necessary and sufficient that $d_1 = 1$ and $d_s = 0$ for $s \neq 1$; these are exactly the conditions (a) and (b). □

Corollary. *Suppose* H *is normal in* G. *In order that* $\mathrm{Ind}_H^G(\rho)$ *be irreducible, it is necessary and sufficient that* ρ *be irreducible and not isomorphic to any of its conjugates* ρ^s *for* $s \notin \mathrm{H}$.

Indeed, we have then $\mathrm{H}_s = \mathrm{H}$ and $\mathrm{Res}_s(\rho) = \rho$.

Exercise

7.4. Let k be a finite field, let $\mathrm{G} = \mathbf{SL}_2(k)$ and let H be the subgroup of G consisting of matrices $\left(\begin{smallmatrix} a & b \\ c & d \end{smallmatrix}\right)$ such that $c = 0$. Let ω be a homomorphism of k^* into $\mathbf{C}^*$ and let χ_ω be the character of degree 1 of H defined by

$$\chi_\omega \begin{pmatrix} a & b \\ 0 & d \end{pmatrix} = \omega(a).$$

Show that the representation of G induced by χ_ω is irreducible if $\omega^2 \neq 1$. Compute χ_ω.

CHAPTER 8

Examples of induced representations

8.1 Normal subgroups; applications to the degrees of the irreducible representations

Proposition 24. *Let* A *be a normal subgroup of a group* G, *and let* $\rho\colon \mathrm{G} \to \mathbf{GL}(\mathrm{V})$ *be an irreducible representation of* G. *Then*:

(a) *either there exists a subgroup* H *of* G, *unequal to* G *and containing* A, *and an irreducible representation* σ *of* H *such that* ρ *is induced by* σ;

(b) *or else the restriction of* ρ *to* A *is isotypic.*

(A representation is said to be *isotypic* if it is a direct sum of isomorphic irreducible representations.)

Let $\mathrm{V} = \oplus \mathrm{V}_i$ be the canonical decomposition of the representation ρ (restricted to A) into a direct sum of isotypic representations (cf. 2.6). For $s \in \mathrm{G}$ we see by "transport de structure" that $\rho(s)$ permutes the V_i; since V is irreducible, G permutes them transitively. Let V_{i_0} be one of these; if V_{i_0} is equal to V, we have case (b). Otherwise, let H be the subgroup of G consisting of those $s \in \mathrm{G}$ such that $\rho(s)\mathrm{V}_{i_0} = \mathrm{V}_{i_0}$. We have $\mathrm{A} \subset \mathrm{H}$, $\mathrm{H} \neq \mathrm{G}$, and ρ is induced by the natural representation σ of H in V_{i_0}, which is case (a). □

Remark. If A is abelian, (b) is equivalent to saying that $\rho(a)$ is a *homothety* for each $a \in \mathrm{A}$.

Corollary. *If* A *is an abelian normal subgroup of* G, *the degree of each irreducible representation* ρ *of* G *divides the index* (G: A) *of* A *in* G.

The proof is by induction on the order of G. In case (a) of the preceding proposition the induction hypothesis shows that the degree of σ divides

(H: A), and by multiplying this relation by (G: H) we see that the degree of ρ divides (G: A). In case (b) let $G' = \rho(G)$ and $A' = \rho(A)$; since the canonical map $G/A \to G'/A'$ is surjective, (G′: A′) divides (G: A). Our previous remark shows now that the elements of A′ are homotheties, thus are contained in the center of G′. By prop. 17 of 6.5, it follows that the degree of ρ divides (G′: A′) and *a fortiori* (G: A). □

Remark. If A is an *abelian subgroup* of G (not necessarily normal) it is no longer true in general that $\deg(\rho)$ divides (G: A), but *nevertheless we have* $\deg(\rho) \leqslant (G: A)$, cf. 3.1, cor. to th. 9.

8.2 Semidirect products by an abelian group

Let A and H be two subgroups of the group G, with A normal. Make the following hypotheses:

(i) A is *abelian*.
(ii) G is the *semidirect product* of H by A.

[Recall that (ii) means that $G = A \cdot H$ and that $A \cap H = \{1\}$, or in other words, that each element of G can be written uniquely as a product ah, with $a \in A$ and $h \in H$.]

We are going to show that the irreducible representations of G can be constructed from those of certain subgroups of H (this is the method of "little groups" of Wigner and Mackey).

Since A is abelian, its irreducible characters are of degree 1 and form a group $X = \mathrm{Hom}(A, \mathbf{C}^*)$. The group G acts on X by

$$(s\chi)(a) = \chi(s^{-1}as) \quad \text{for } s \in G, \chi \in X, a \in A.$$

Let $(\chi_i)_{i \in X/H}$ be a system of representatives for the orbits of H in X. For each $i \in X/H$, let H_i be the subgroup of H consisting of those elements h such that $h\chi_i = \chi_i$, and let $G_i = A \cdot H_i$ be the corresponding subgroup of G. Extend the function χ_i to G_i by setting

$$\chi_i(ah) = \chi_i(a) \quad \text{for } a \in A, h \in H_i.$$

Using the fact that $h\chi_i = \chi_i$ for all $h \in H_i$, we see that χ_i is a *character of degree* 1 of G_i. Now let ρ be an irreducible representation of H_i; by composing ρ with the canonical projection $G_i \to H_i$ we obtain an irreducible representation $\tilde{\rho}$ of G_i. Finally, by taking the tensor product of χ_i and $\tilde{\rho}$ we obtain an irreducible representation $\chi_i \otimes \tilde{\rho}$ of G_i; let $\theta_{i,\rho}$ be the corresponding induced representation of G.

Proposition 25

(a) *$\theta_{i,\rho}$ is irreducible.*
(b) *If $\theta_{i,\rho}$ and $\theta_{i',\rho'}$ are isomorphic, then $i = i'$ and ρ is isomorphic to ρ'.*
(c) *Every irreducible representation of* G *is isomorphic to one of the $\theta_{i,\rho}$.*

(Thus we have all the irreducible representations of G.)

We prove (a) using *Mackey's criterion* (7.4, prop. 23) as follows: Let $s \notin G_i = A \cdot H_i$, and let $K_s = G_i \cap sG_is^{-1}$. We have to show that, if we compose the representation $\chi_i \otimes \tilde{\rho}$ of G_i with the two injections $K_s \to G_i$ defined by $x \mapsto x$ and $x \mapsto s^{-1}xs$, we obtain two disjoint representations of K_s. To do this, it is enough to check that the restrictions of these representations to the subgroup A of K_s are disjoint. But the first restricts to a multiple of χ_i and the second to a multiple of $s\chi_i$; since $s \notin A \cdot H_i$ we have $s\chi_i \neq \chi_i$ and so the two representations in question are indeed disjoint.

Now we prove (b). First of all, the restriction of $\theta_{i,\rho}$ to A only involves characters χ belonging to the orbit $H\chi_i$ of χ_i. This shows that $\theta_{i,\rho}$ determines i. Next, let W be the representation space for $\theta_{i,\rho}$, and let W_i be the subspace of W corresponding to χ_i [i.e., the set of $x \in W$ such that $\theta_{i,\rho}(a)x = \chi_i(a)x$ for all $a \in A$]. The subspace W_i is stable under H_i, and one checks immediately that the representation of H_i in W_i is isomorphic to ρ; whence $\theta_{i,\rho}$ determines ρ.

Finally, let $\sigma\colon G \to \mathbf{GL}(W)$ be an irreducible representation of G. Let $W = \bigoplus_{\chi \in X} W$ be the canonical decomposition of $\mathrm{Res}_A W$. At least one of the W_χ is nonzero; if $s \in G$, $\sigma(s)$ transforms W_χ into $W_{s(\chi)}$. The group H_i maps W_{χ_i} into itself; let W_i be an irreducible sub - $\mathbf{C}[H_i]$-module of W_{χ_i} and let ρ be the corresponding representation of H_i. It is clear that the representation of $G_i = A \cdot H_i$ is isomorphic to $\chi_i \otimes \tilde{\rho}$. Thus the restriction of σ to G_i contains $\chi_i \otimes \tilde{\rho}$ at least once. By prop. 21, it follows that σ occurs at least once in the induced representation $\theta_{i,\rho}$; since $\theta_{i,\rho}$ is irreducible, this implies that σ and $\theta_{i,\rho}$ are isomorphic, which proves (c). □

EXERCISES

8.1. Let a, h, h_i be the orders of A, H, H_i respectively. Show that $a = \sum (h/h_i)$. Show that, for fixed i, the sum of the squares of the degrees of the representations $\theta_{i,\rho}$ is h^2/h_i. Deduce from this another proof of (c).

8.2. Use prop. 25 to recompute the irreducible representations of the groups D_n, $\mathfrak{A}_4$, and $\mathfrak{S}_4$ (cf. Ch. 5).

8.3 A review of some classes of finite groups

For more details on the results of this section and the following, see Bourbaki, Alg. I, §7.

Solvable groups. One says that G is solvable if there exists a sequence

$$\{1\} = G_0 \subset G_1 \subset \cdots \subset G_n = G$$

of subgroups of G, with G_{i-1} normal in G_i and G_i/G_{i-1} abelian for

$1 \leqslant i \leqslant n$. (Equivalent definition: G is obtained from the group $\{1\}$ by a finite number of *extensions with abelian kernels*.)

Supersolvable groups. Same as above, except that one requires that all the G_i be normal in G and that G_i/G_{i-1} be cyclic.

Nilpotent groups. As above, except that G_i/G_{i-1} is required to be in the center of G/G_{i-1} for $1 \leqslant i \leqslant n$. (Equivalent definition: G is obtained from the group $\{1\}$ by a finite number of *central extensions*.)

It is clear that supersolvable $\Rightarrow$ solvable. On the other hand, one checks immediately that each central extension of a supersolvable group is supersolvable; thus nilpotent $\Rightarrow$ supersolvable.

p-groups. If p is a prime, a group whose order is a power of p is called a p-group.

Theorem 14. *Every p-group is nilpotent (thus supersolvable).*

In view of the preceding it suffices to show that the center of every nontrivial p-group G is nontrivial. This is a consequence of the following lemma:

Lemma 3. *Let* G *be a p-group acting on a finite set* X, *and let* X^G *be the set of elements of* X *fixed by* G. *We have*

$$\mathrm{Card}(X) \equiv \mathrm{Card}(X^G) \qquad (\mathrm{mod.}\, p).$$

Indeed $X - X^G$ is a union of nontrivial orbits of G, and the cardinality of each of these orbits is a power p^α of p, with $\alpha \geqslant 1$; hence $\mathrm{Card}(X - X^G)$ is divisible by p. □

Let us now apply this lemma to the case $X = G$ with G acting by inner automorphisms. The set X^G is just the *center* C of G. Thus

$$\mathrm{Card}(C) \equiv \mathrm{Card}(G) \equiv 0 \ (\mathrm{mod.}\, p),$$

whence $C \neq \{1\}$, which proves the theorem.

We record another application of lemma 3 which will be used in Part III:

Proposition 26. *Let* V *be a vector space* $\neq 0$ *over a field k of characteristic p and let* $\rho\colon G \to \mathbf{GL}(V)$ *be a linear representation of a p-group* G *in* V. *Then there exists a nonzero element of* V *which is fixed by all* $\rho(s)$, $s \in G$.

Let x be a nonzero element of V, and let X be the subgroup of V generated by the $\rho(s)x$, $s \in G$. We apply lemma 3 to X, observing that X is finite and of order a power of p. Therefore $X^G \neq \{0\}$, which proves the proposition. □

Corollary. *The only irreducible representation of a p-group in characteristic p is the trival representation.*

EXERCISES

8.3. Show that the dihedral group D_n is supersolvable, and that it is nilpotent if and only if n is a power of 2.

8.4. Show that the alternating group $\mathfrak{A}_4$ is solvable, but not supersolvable. Same question for the group $\mathfrak{S}_4$.

8.5. Show that each subgroup and each quotient of a solvable group (resp. supersolvable, nilpotent) is solvable (resp. supersolvable, nilpotent).

8.6. Let p and q be distinct prime numbers and let G be a group of order $p^a q^b$ where a and b are integers > 0.

(i) Assume that the center of G is $\{1\}$. For $s \in$ G denote by $c(s)$ the number of elements in the conjugacy class of s. Show that there exists $s \neq 1$ such that $c(s) \not\equiv 0$ (mod.q). (Otherwise the number of elements of G $- \{1\}$ would be divisible by q.) For such an s, $c(s)$ is a power of p; derive from this the existence of a normal subgroup of G unequal to $\{1\}$ or G [apply ex. 6.10].

(ii) Show that G is solvable (*Burnside's theorem*). [Use induction on the order of G and distinguish two cases, depending on whether the center of G is equal or unequal to $\{1\}$.]

(iii) Show by example that G is not necessarily supersolvable (cf. ex. 8.4).

(iv) Give an example of a nonsolvable group whose order is divisible by just three prime numbers [$\mathfrak{S}_5$, $\mathfrak{S}_6$, $\mathbf{GL}_2(\mathbf{F}_7)$ will do].

8.4 Sylow's theorem

Let p be a prime number, and let G be a group of order $g = p^n m$, where m is prime to p. A subgroup of G of order p^n is called a *Sylow p-subgroup* of G.

Theorem 15

(a) *There exist Sylow p-subgroups.*
(b) *They are conjugate by inner automorphisms.*
(c) *Each p-subgroup of* G *is contained in a Sylow p-subgroup.*

To prove (a) we use induction on the order of G. We may assume $n \geqslant 1$, i.e. Card (G) $\equiv 0$ (mod. p). Let C be the center of G. If Card (C) is divisible of order p, an elementary argument shows that C contains a subgroup D cyclic of order p. By the induction hypothesis, G/D has a Sylow p-subgroup, and the inverse image of this subgroup in G is a Sylow p-subgroup of G. If Card (C) $\not\equiv 0$ (mod. p) let G act on G $-$ C by inner automorphisms; this gives a partition of $\mathrm{G} - \mathrm{C}$ into orbits (conjugacy classes). As $\mathrm{Card}(\mathrm{G} - \mathrm{C}) \not\equiv 0 \pmod{p}$, one of these orbits has a cardinality prime to p. It follows that there is a subgroup H unequal to G such that $(\mathrm{G}: \mathrm{H}) \not\equiv 0 \pmod{p}$. The order of H is thus divisible by p^n, and the induction hypothesis shows that H contains a subgroup of order p^n.

Now let P be a Sylow p-subgroup of G and Q a p-subgroup of G. The p-group Q acts on $\mathrm{X} = \mathrm{G}/\mathrm{P}$ by left translations. By lemma 3 of 8.3 we have

$$\mathrm{Card}(\mathrm{X}^{\mathrm{Q}}) \equiv \mathrm{Card}(\mathrm{X}) \not\equiv 0 \pmod{p},$$

whence $\mathrm{X}^{\mathrm{Q}} \neq \varnothing$. Thus there exists an element $x \in \mathrm{G}$ such that $\mathrm{Q}x\mathrm{P} = x\mathrm{P}$, hence $\mathrm{Q} \subset x\mathrm{P}x^{-1}$, which proves (c). If in addition $\mathrm{Card}(\mathrm{Q}) = p^n$, the groups Q and $x\mathrm{P}x^{-1}$ have the same order, and $\mathrm{Q} = x\mathrm{P}x^{-1}$, which proves (b). ☐

EXERCISES

8.7. Let H be a normal subgroup of a group G and let $\mathrm{P_H}$ be a Sylow p-subgroup of G/H.

(a) Show that there exists a Sylow p-subgroup P of G whose image in G/H is $\mathrm{P_H}$ [use the conjugacy of Sylow subgroups].

(b) Show that P is unique if H is a p-group or if H is in the center of G [reduce to the case where H has order prime to p, and use the fact that each homomorphism from $\mathrm{P_H}$ into H is trivial].

8.8. Let G be a nilpotent group. Show that, for each prime number p, G contains a unique Sylow p-subgroup, which is normal [use induction on the order of G, and apply the induction hypothesis to the quotient of G by its center, cf. ex. 8.7(b)]. Conclude that G is a direct product of p-groups.

8.9. Let $\mathrm{G} = \mathbf{GL}_n(k)$, where k is a finite field of characteristic p. Show that the subgroup of G which consists of all upper triangular matrices having only 1's on the diagonal is a Sylow p-subgroup of G.

8.5 Linear representations of supersolvable groups

Lemma 4. *Let* G *be a nonabelian supersolvable group. Then there exists a normal abelian subgroup of* G *which is not contained in the center of* G.

Let C be the center of G. The quotient $\mathrm{H} = \mathrm{G}/\mathrm{C}$ is supersolvable, thus has a composition series in which the first nontrivial term H_1 is a cyclic normal subgroup of H. The inverse image of H_1 in G has the required properties. ☐

Theorem 16. *Let* G *be a supersolvable group. Then each irreducible representation of* G *is induced by a representation of degree* 1 *of a subgroup of* G *(i.e., is monomial).*

We prove the theorem by induction on the order of G. Consequently we may consider only those irreducible representations ρ which are *faithful*, i.e., such that $\mathrm{Ker}(\rho) = \{1\}$. If G is abelian, such a ρ is of degree 1 and there is nothing to prove. Suppose G is not abelian, and let A be a normal abelian subgroup of G which is not contained in the center of G (cf. lemma 4). Since ρ is faithful, this implies that $\rho(\mathrm{A})$ is not contained in the center of $\rho(\mathrm{G})$; thus there exists $a \in \mathrm{A}$ such that $\rho(a)$ is not a homothety. The restriction of ρ to A is thus not isotypic. By prop. 24, this implies that ρ is induced by an irreducible representation of a subgroup H of G which is unequal to G. The theorem now follows by applying induction to H. □

Exercises

8.10. Extend Theorem 16 to groups which are semidirect products of a supersolvable group by an abelian normal subgroup [use prop. 25 to reduce to the supersolvable case].

8.11. Let **H** be the field of quaternions over **R**, with basis $\{1, i, j, k\}$ satisfying

$$i^2 = j^2 = k^2 = -1, \qquad ij = -ji = k, \qquad jk = -kj = i,$$
$$ki = -ik = j.$$

Let E be the subgroup of $\mathbf{H}^*$ consisting of the eight elements ± 1, $\pm i$, $\pm j$, $\pm k$ (quaternion group), and let G be the union of E and the sixteen elements $(\pm 1 \pm i \pm j \pm k)/2$. Show that G is a solvable subgroup of $\mathbf{H}^*$ which is a semidirect product of a cyclic group of order 3 by the normal subgroup E. Use the isomorphism $\mathbf{H} \otimes_{\mathbf{R}} \mathbf{C} = \mathbf{M}_2(\mathbf{C})$ to define an irreducible representation of degree 2 of G. Show that this representation is not monomial (observe that G has no subgroup of index 2). [The group G is the group of invertible elements of the ring of Hurwitz "integral quaternions"; it is also the group of automorphisms of the elliptic curve $y^2 - y = x^3$ in characteristic 2. It is isomorphic to $\mathbf{SL}_2(\mathbf{F}_3)$.]

8.12. Let G be a p-group. Show that, for each irreducible character χ of G, we have $\sum \chi'(1)^2 \equiv 0 \pmod{\chi(1)^2}$, the sum being over all irreducible characters χ' such that $\chi'(1) < \chi(1)$. [Use the fact that $\chi(1)$ is a power of p, and apply cor. 2(a) to prop. 5.]

CHAPTER 9

Artin's theorem

9.1 The ring R(G)

Let G be a finite group and let $\chi_1, \ldots, \chi_h$ be its distinct irreducible characters. A class function on G is a character if and only if it is a linear combination of the χ_i's with non-negative integer coefficients. We will denote by $R^+(G)$ the set of these functions, and by R(G) the group generated by $R^+(G)$, i.e., the set of differences of two characters. We have

$$R(G) = \mathbf{Z}\chi_1 \oplus \cdots \oplus \mathbf{Z}\chi_h.$$

An element of R(G) is called a *virtual character*. Since the product of two characters is a character, R(G) is a *subring* of the ring $F_{\mathbf{C}}(G)$ of class functions on G with complex values. Since the χ_i form a basis of $F_{\mathbf{C}}(G)$ over $\mathbf{C}$, we see that $\mathbf{C} \otimes R(G)$ *can be identified with* $F_{\mathbf{C}}(G)$.

We can also view R(G) as the *Grothendieck group* of the category of finitely generated $\mathbf{C}[G]$-modules; this will be used in Part III.

If H is a subgroup of G, the operation of *restriction* defines a ring homomorphism $R(G) \to R(H)$, denoted by Res_H^G or Res.

Similarly, the operation of induction (7.2) defines a homomorphism of abelian groups $R(H) \to R(G)$, denoted by Ind_H^G or Ind. The homomorphisms Ind and Res are adjoints of each other with respect to the bilinear forms $\langle\varphi, \psi\rangle_H$ and $\langle\varphi, \psi\rangle_G$, cf. th. 13. Moreover, the formula

$$\mathrm{Ind}(\varphi \cdot \mathrm{Res}(\psi)) = \mathrm{Ind}(\varphi) \cdot \psi$$

shows that the image of Ind: $R(H) \to R(G)$ is an ideal of the ring R(G).

If A is a commutative ring, the homomorphisms Res and Ind extend by linearity to A-linear maps:

$$\mathrm{A} \otimes \mathrm{Res}\colon \mathrm{A} \otimes \mathrm{R(G)} \to \mathrm{A} \otimes \mathrm{R(H)}$$

$$\mathrm{A} \otimes \mathrm{Ind}\colon \mathrm{A} \otimes \mathrm{R(H)} \to \mathrm{A} \otimes \mathrm{R(G)}$$

Exercises

9.1. Let φ be a real-valued class function on G. Assume that $\langle \varphi, 1\rangle = 0$ and that $\varphi(s) \leqslant 0$ for each $s \neq 1$. Show that for each character χ the real part of $\langle \varphi, \chi\rangle$ is $\geqslant 0$ [use the fact that the real part of $\varphi(s^{-1})\chi(s)$ is greater than or equal to that of $\varphi(s^{-1})\chi(1)$ for all s]. Conclude that, if φ belongs to R(G), φ is a character.

9.2. Let $\chi \in \mathrm{R(G)}$. Show that χ is an irreducible character if and only if $\langle \chi, \chi\rangle = 1$ and $\chi(1) \geqslant 0$.

9.3. If f is a function on G, and k an integer, denote by $\Psi^k(f)$ the function $s \mapsto f(s^k)$.

(a) Let ρ be a representation of G with character χ. For each integer $k \geqslant 0$, denote by χ_σ^k (resp. χ_λ^k) the character of the kth *symmetric power* (resp. kth *exterior power*) *of* ρ (cf. 2.1 for the case $k = 2$). Set

$$\sigma_{\mathrm{T}}(\chi) = \sum_{k=0}^{\infty} \chi_\sigma^k \mathrm{T}^k \quad \text{and} \quad \lambda_{\mathrm{T}}(\chi) = \sum_{k=0}^{\infty} \chi_\lambda^k \mathrm{T}^k,$$

where T is an indeterminate. Show that, for $s \in \mathrm{G}$, we have

$$\sigma_{\mathrm{T}}(\chi)(s) = 1/\det(1 - \rho(s)\mathrm{T}) \quad \text{and} \quad \lambda_{\mathrm{T}}(\chi)(s) = \det(1 + \rho(s)\mathrm{T}).$$

Deduce the formulas

$$\sigma_{\mathrm{T}}(\chi) = \exp\left\{\sum_{k=1}^{\infty} \Psi^k(\chi)\mathrm{T}^k/k\right\},$$

$$\lambda_{\mathrm{T}}(\chi) = \exp\left\{\sum_{k=1}^{\infty} (-1)^{k-1}\Psi^k(\chi)\mathrm{T}^k/k\right\},$$

and

$$n\chi_\sigma^n = \sum_{k=1}^{n} \Psi^k(\chi)\chi_\sigma^{n-k}, \qquad n\chi_\lambda^n = \sum_{k=1}^{n} (-1)^{k-1}\Psi^k(\chi)\chi_\lambda^{n-k},$$

which generalize those of 2.1.

(b) Conclude from a) that R(G) is *stable under the operators* Ψ^k, $k \in \mathbf{Z}$.

9.4. Let n be an integer prime to the order of G.

(a) Let χ be an irreducible character of G. Show that $\Psi^n(\chi)$ is an irreducible character of G [use the two preceding exercises].

(b) Extend by linearity $x \mapsto x^n$ to an endomorphism ψ_n of the vector space $\mathbf{C}[G]$. Show that the restriction of ψ_n to Cent. $\mathbf{C}[G]$ is an *automorphism* of the algebra Cent. $\mathbf{C}[G]$.

9.2 Statement of Artin's theorem

It is as follows:

Theorem 17. *Let* X *be a family of subgroups of a finite group* G. *Let* Ind: $\oplus_{H \in X} R(H) \to R(G)$ *be the homomorphism defined by the family of* Ind_H^G, $H \in X$. *Then the following properties are equivalent*:

(i) G *is the union of the conjugates of the subgroups belonging to* X.
(ii) *The cokernel of* Ind: $\bigoplus_{H \in X} R(H) \to R(G)$ *is finite.*

Since R(G) is finitely generated as a group, we can rephrase (ii) in the following way:

(ii′) *For each character* χ *of* G, *there exist virtual characters* $\chi_H \in R(H)$, $H \in X$, *and an integer* $d \geqslant 1$ *such that*

$$d\chi = \sum_{H \in X} \mathrm{Ind}_H^G(\chi_H).$$

Note that the family of cyclic subgroups of G satisfies (i). Hence:

Corollary. *Each character of* G *is a linear combination with rational coefficients of characters induced by characters of cyclic subgroups of* G.

We will see in the next section that the above statement remains true when "rational" is replaced by "integer" and "cyclic" by "elementary."

EXERCISE

9.5. Take for G the alternating group $\mathfrak{A}_4$ and for X the family of cyclic subgroups of G. Let $\{\chi_0, \chi_1, \chi_2, \psi\}$ be the distinct irreducible characters of G (cf. 5.7). Show that the image of $\bigoplus_{H \in X} R^+(H)$ under Ind is generated by the five characters:

$$\chi_0 + \chi_1 + \chi_2 + \psi, \qquad 2\psi, \qquad \chi_0 + \psi, \qquad \chi_1 + \psi, \qquad \chi_2 + \psi.$$

Conclude that an element χ of R(G) belongs to the image of Ind if and only if $\chi(1) \equiv 0 \pmod{2}$. Show that none of the characters χ_0, χ_1, χ_2 is a linear combination with *positive* rational coefficients of characters induced from cyclic subgroups.

9.3 First proof

First, we show that (ii) ⇒ (i). Let S be the union of the conjugates of the subgroups H belonging to X. Each function of the form $\sum \mathrm{Ind}_H^G(f_H)$, with $f_H \in R(H)$, vanishes off S. If (ii) is satisfied, it follows that each class function on G vanishes off S, which shows that S = G. Hence (i) holds.

Conversely, suppose (i) is satisfied. To prove (ii), it suffices to show that the **Q**-linear map

$$\mathbf{Q} \otimes \mathrm{Ind}\colon \bigoplus_{H \in X} \mathbf{Q} \otimes R(H) \to \mathbf{Q} \otimes R(G).$$

is surjective, which is also equivalent to the surjectivity of the **C**-linear map

$$\mathbf{C} \otimes \mathrm{Ind}\colon \bigoplus_{H \in X} \mathbf{C} \otimes R(H) \to \mathbf{C} \otimes R(G)$$

By duality this is equivalent to the *injectivity* of the adjoint map

$$\mathbf{C} \otimes \mathrm{Res}\colon \mathbf{C} \otimes R(G) \to \bigoplus_{H \in X} \mathbf{C} \otimes R(H).$$

But this injectivity is obvious: it amounts to saying that if a class function on G restricts to 0 on each cyclic subgroup, then it is zero. The theorem follows. □

Exercises

We assume that the family X is stable under conjugation and passage to subgroups, and that G is the union of the subgroups belonging to X. (Example: the family of cyclic subgroups of G.)

9.6. Denote by N the kernel of the homomorphism

$$\mathbf{Q} \otimes \mathrm{Ind}\colon \bigoplus_{H \in X} \mathbf{Q} \otimes R(H) \to \mathbf{Q} \otimes R(G).$$

(a) Let H, $H' \in X$, with $H' \subset H$, let $\chi' \in R(H')$ and $\chi = \mathrm{Ind}_{H'}^{H}(\chi') \in R(H)$. Show that $\chi - \chi'$ belongs to N.

(b) Let $H \in X$ and $s \in G$. Set ${}^sH = sHs^{-1}$. Let $\chi \in R(H)$ and let ${}^s\chi$ be the element of $R({}^sH)$ defined by ${}^s\chi(shs^{-1}) = \chi(h)$ for $h \in H$. Show that $\chi - {}^s\chi$ belongs to N.

(c) Show that N is generated over **Q** by the elements of type (a) and (b) above. [Extend scalars to **C** and use duality. One is led to prove that, if for each $H \in X$ a class function f_H on H is given and if the f_H satisfy conditions of restriction and conjugation analogous to (a) and (b) above, then there exists a class function f on G such that $\mathrm{Res}_H^G f = f_H$ for each H.]

9.7. Show that $\mathbf{Q} \otimes R(G)$ has a presentation* by generators and relations of the following form:

Generators: symbols (H, χ), with $H \in X$ and $\chi \in \mathbf{Q} \otimes R(H)$.

---

* This exercise gives a "presentation" of $\mathbf{Q} \otimes R(G)$ in terms of induced characters (H, χ). It would be very desirable, for application to the theory of L-series, to give such a presentation for R(G) itself (without tensoring by **Q**). When G is solvable, this has been done by Langlands-Deligne (Lecture Notes in Math. 349, p. 517, th. 4).

Relations:

(i) $(H, \lambda\chi + \lambda'\chi') = \lambda(H,\chi) + \lambda'(H,\chi')$ for $\lambda, \lambda' \in \mathbf{Q}$, and $\chi, \chi' \in \mathbf{Q} \otimes R(H)$.

(ii) For $H' \subset H$, $\chi' \in R(H')$, and $\chi = \mathrm{Ind}_{H'}^{H}(\chi')$, we have $(H,\chi) = (H',\chi')$.

(iii) For $H \in X$, $s \in G$, $\chi \in R(H)$, we have $(H,\chi) = ({}^{s}H, {}^{s}\chi)$, with the notation of ex. 9.6(b).

[Use ex. 9.6].

9.4 Second proof of (i) ⇒ (ii)

First let A be a cyclic group, and let a be its order. Define a function θ_A on A by the formula:

$$\theta_A(x) = \begin{cases} a & \text{if } x \text{ generates A} \\ 0 & \text{otherwise} \end{cases}$$

Proposition 27. *If* G *is a finite group of order* g, *then*

$$g = \sum_{A \subset G} \mathrm{Ind}_A^G(\theta_A),$$

where A *runs through all the cyclic subgroups of* G.

(In this formula, the letter g denotes the constant function equal to g.)

Put $\theta'_A = \mathrm{Ind}_A^G(\theta_A)$. For $x \in G$ we have

$$\theta'_A(x) = \frac{1}{a} \sum_{\substack{y \in G \\ yxy^{-1} \in A}} \theta_A(yxy^{-1})$$

$$= \frac{1}{a} \sum_{\substack{y \in G \\ yxy^{-1}\,\mathrm{gen.}\,A}} a = \sum_{\substack{y \in G \\ yxy^{-1}\,\mathrm{gen.}\,A}} 1.$$

However, for each $y \in G$, yxy^{-1} generates a unique cyclic subgroup of G. So we have:

$$\sum_{A \subset G} \theta'_A(x) = \sum_{y \in G} 1 = g. \qquad \square$$

Proposition 28. *If* A *is a cyclic group, then* $\theta_A \in R(A)$.

The proof is by induction on the order a of A, the case $a = 1$ being trivial. By prop. 27 we have

$$a = \sum_{B \subset A} \mathrm{Ind}_B^A(\theta_B) = \theta_A + \sum_{B \neq A} \mathrm{Ind}_B^A(\theta_B).$$

The induction hypothesis gives $\theta_B \in R(B)$ for $B \neq A$, hence $\mathrm{Ind}_B^A(\theta_B)$ belongs to $R(A)$; on the other hand, it is clear that $a \in R(A)$ and so it follows that θ_A belongs to $R(A)$. □

Application to the proof of (i) ⇒ (ii)

First observe that, if A' is contained in a conjugate of A, the image of $\mathrm{Ind}_{A'}^G$ is contained in that of Ind_A^G. Hence we can assume that X is the family of all cyclic subgroups of G. Propositions 27 and 28 then show that

$$g = \sum_{A \in X} \mathrm{Ind}_A^G(\theta_A), \qquad \text{with } \theta_A \in R(A).$$

Thus the element g belongs to the image of Ind. Since this image is an *ideal* of R(G), cf. 9.1, it contains every element of the form $g\chi$, with $\chi \in R(G)$, which proves (ii′) (and even more, since we have an explicit denominator *viz.* the order of G).

Exercise

9.8. If A is cyclic of order a, put $\lambda_A = \varphi(a)r_A - \theta_A$, where $\varphi(a)$ is the number of generators of A, and r_a is the character of the regular representation. Show that λ_A is a character of A orthogonal to the unit character [apply ex. 9.1]. Show that, if A runs over the set of cyclic subgroups of a group G of order g, we have

$$(*) \qquad \sum_{A \subset G} \mathrm{Ind}_A^G(\lambda_A) = g(r_G - 1),$$

where r_G is the character of the regular representation of G [use prop. 27].

[Application (Aramata–Brauer): Let F be a finite extension of the number field E, and let $\Phi(s) = \zeta_F(s)/\zeta_E(s)$ be the quotient of their zeta functions. It is known that Φ is meromorphic in the entire complex plane. Now suppose that F/E is a Galois extension with Galois group G. Then the formula $(*)$ above implies the identity

$$\Phi(s)^g = \prod_A L_{F/F_A}(s, \lambda_A),$$

where F_A denotes the subfield of F corresponding to the cyclic subgroup A. The functions $L_{F/F_A}(s, \lambda_A)$ are "abelian" L-functions, and hence holomorphic. So we see that Φ itself is holomorphic, i.e., that ζ_E *divides* ζ_F; it is not known if this result still holds for non-Galois extensions (this would follow from conjectures of Artin).]

CHAPTER 10

A theorem of Brauer

In sections 10.1 through 10.4 the letter p denotes a prime number.

10.1 p-regular elements; p-elementary subgroups

Let x be an element of a finite group G. We say that x is a *p-element* (or is *p-unipotent*) if x has order a power of p; we say that x is a *p'-element* (or is *p-regular*) if its order is prime to p.

Each $x \in G$ can be written in a unique way $x = x_u x_r$ where x_u is p-unipotent, x_r is p-regular, and x_u and x_r commute; moreover, x_u and x_r are powers of x. This can be seen by decomposing the cyclic subgroup generated by x as a direct product of its p-component and its p'-component. The element x_u (resp. x_r) is called the *p-component* (resp. the *p'-component*) of x.

A group H is said to be *p-elementary* if it is the direct product of a cyclic group C of order prime to p with a p-group P. Such a group is *nilpotent* and its decomposition $C \times P$ is unique: C is the set of p'-elements of H, and P is the set of p-elements.

Let x be a p'-element of a finite group G, let C be the cyclic subgroup generated by x, and let $Z(x)$ be the centralizer of x (the set of all $s \in G$ such that $sx = xs$). If P is a Sylow p-subgroup of $Z(x)$, the group $H = C \cdot P$ is a p-elementary subgroup of G, which is said to be *associated* with x; it is unique up to conjugation in $Z(x)$.

EXERCISES

10.1. Let $H = C \cdot P$ be a p-elementary subgroup of a finite group G, and let x be a generator of C. Show that H is contained in a p-elementary subgroup H′ associated with x.

10.2. Let $G = \mathbf{GL}_n(k)$, where k is a finite field of characteristic p. Show that an element $x \in G$ is a p-element if and only if its eigenvalues are all equal to 1, i.e., if $1 - x$ is nilpotent; it is a p'-element if and only if it is semisimple, i.e., diagonalizable in a finite extension of k.

10.2 Induced characters arising from p-elementary subgroups

The purpose of this and the next two sections is to prove the following result:

Theorem 18. *Let* G *be a finite group and let* V_p *be the subgroup of* R(G) *generated by characters induced from those of p-elementary subgroups of* G. *Then the index of* V_p *in* R(G) *is finite and prime to p.*

Let $X(p)$ be the family of p-elementary subgroups of G. The group V_p is the image of the homomorphism

$$\text{Ind}: \bigoplus_{H \in X(p)} R(H) \to R(G)$$

defined by the induction homomorphisms Ind_H^G, $H \in X(p)$. Then V_p is an ideal of R(G), and to prove the theorem it is enough to show that there exists an integer m, prime to p, such that $m \in V_p$. In fact, we prove the following more precise result:

Theorem 18′. *Let* $g = p^n l$ *be the order of* G, *with* $(p, l) = 1$. *Then* $l \in V_p$.

The proof (due to Roquette and Brauer-Tate [12]) uses the subring A of $\mathbf{C}$ generated by the gth roots of unity. This ring is free and finitely generated as a $\mathbf{Z}$-module; its elements are algebraic integers. We have $\mathbf{Q} \cap A = \mathbf{Z}$, since the elements of this intersection are simultaneously rational numbers and algebraic integers (cf. 6.4). The quotient group $A/\mathbf{Z}$ is finitely generated and torsion-free, hence free; it follows (by lifting to A a basis of $A/\mathbf{Z}$) that A has a basis $\{1, \alpha_1, \ldots, \alpha_c\}$ containing the element 1.

The homomorphism Ind defines, by tensoring with A, an A-linear map

$$A \otimes \text{Ind}: \bigoplus_{H \in X(p)} A \otimes R(H) \to A \otimes R(G).$$

The existence of the basis $\{1, \alpha_1, \ldots, \alpha_c\}$ then implies the following:

Lemma 5. *The image of* $A \otimes \text{Ind}$ *is* $A \otimes V_p$; *moreover we have*

$$(A \otimes V_p) \cap R(G) = V_p.$$

Thus, to prove that the constant function l belongs to V_p, it is enough to prove that l belongs to the image of $A \otimes \text{Ind}$, or in other words, that l is of the form $\sum_H a_H \text{Ind}_H^G(f_H)$, with $a_H \in A$ and $f_H \in R(H)$.

Remarks

(1) The advantage of the ring A over the ring **Z** is that all the characters of G have *values in* A, since these values are sums of gth roots of unity. It follows that $A \otimes R(G)$ is a *subring* of the ring of class functions on G with values in A.

(2) It can be shown that A is the set of algebraic integers of the cyclotomic field $\mathbf{Q} \cdot A$, but we will not need this.

10.3 Construction of characters

Lemma 6. *Each class function on* G *with integer values divisible by g is an* A-*linear combination of characters induced from characters of cyclic subgroups of* G.

(Here, and in all that follows, the expression "integer values" means "values in **Z**.")

Let f be such a function, and write it in the form $g\chi$, where χ is a class function with integer values. If C is a cyclic subgroup of G, let θ_C be the element of R(C) defined in 9.4. We have

$$g = \sum_C \mathrm{Ind}_C^G(\theta_C), \quad \text{cf. prop. 27,}$$

whence

$$f = g\chi = \sum_C \mathrm{Ind}_C^G(\theta_C)\chi = \sum_C \mathrm{Ind}_C^G(\theta_C \cdot \mathrm{Res}_C^G \chi).$$

It remains to show that $\theta_C . \mathrm{Res}_C\, \chi$ belongs to $A \otimes R(C)$ for each C. But the the values of $\chi_C = \theta_C \cdot \mathrm{Res}_C \chi$ are divisible by the order of C, so if ψ is a character of C, we have $\langle \chi_C, \psi \rangle \in A$, which shows that χ_C is an A-linear combination of characters of C, whence $\chi_C \in A \otimes R(C)$. □

Lemma 7. *Let χ be an element of* $A \otimes R(G)$ *with integer values, let $x \in G$, and let x_r be the p'-component of x* (cf. 10.1). *Then*

$$\chi(x) \equiv \chi(x_r) \pmod{p}.$$

By restriction, we are led to the case where G is cyclic and generated by x. Now $\chi = \sum a_i \chi_i$, with $a_i \in A$ and the χ_i running over the distinct characters of degree 1 of G. If q is a sufficiently large power of p, we have $x^q = x_r^q$ and thus $\chi_i(x)^q = \chi_i(x_r)^q$ for all i. Hence

$$\begin{aligned} \chi(x)^q &= (\textstyle\sum a_i \chi_i(x))^q \equiv \sum a_i^q \chi_i(x)^q \\ &\equiv \textstyle\sum a_i^q \chi_i(x_r)^q) \equiv \chi(x_r)^q \pmod{p\,A}. \end{aligned}$$

Since $pA \cap \mathbf{Z} = p\mathbf{Z}$, this implies

$$\chi(x)^q \equiv \chi(x_r)^q \pmod{p},$$

hence $\chi(x) \equiv \chi(x_r) \pmod{p}$, since $\lambda^q = \lambda \pmod{p}$ for all $\lambda \in \mathbf{Z}$. □

Lemma 8. *Let x be a p'-element of* G, *and let* H *be a p-elementary subgroup of* G *associated with x* (10.1). *Then there exists a function $\psi \in \mathrm{A} \otimes \mathrm{R(H)}$, with integer values, such that the induced function $\psi' = \mathrm{Ind}_{\mathrm{H}}^{\mathrm{G}}\psi$ has the following properties*:

(a) $\psi'(x) \not\equiv 0 \pmod{p}$.
(b) $\psi'(s) = 0$ *for each p'-element of* G *which is not conjugate to x.*

Let C be the cyclic subgroup of G generated by x, and let $\mathrm{Z}(x)$ be the centralizer of x in G. We have $\mathrm{H} = \mathrm{C} \times \mathrm{P}$, where P is a Sylow p-subgroup of $\mathrm{Z}(x)$. Let c be the order of C, and let p^a be the order of P. Let ψ_{C} be the function defined on C by

$$\psi_{\mathrm{C}}(x) = c \qquad \text{and} \qquad \psi_{\mathrm{C}}(y) = 0 \quad \text{if } y \neq x.$$

We have $\psi_{\mathrm{C}} = \sum_\chi \chi(x^{-1})\chi$, where χ runs through the set of irreducible characters of C; it follows that ψ_{C} belongs to $\mathrm{A} \otimes \mathrm{R(C)}$ (which follows also from lemma 6).

Let ψ be the function on $\mathrm{H} = \mathrm{C} \times \mathrm{P}$ defined by $\psi(xy) = \psi_{\mathrm{C}}(x)$ for $x \in \mathrm{C}$ and $y \in \mathrm{P}$. This is the inverse image of ψ_{C} under the projection $\mathrm{H} \to \mathrm{C}$. So we have $\psi \in \mathrm{A} \otimes \mathrm{R(H)}$. We show now that ψ satisfies the conditions of the lemma:

If s is a p'-element of G and if $y \in \mathrm{G}$, ysy^{-1} is a p'-element; if ysy^{-1} belongs to H then it belongs to C, and we have $\psi(ysy^{-1}) = 0$ whenever $ysy^{-1} \neq x$. It follows that $\psi'(s) = 0$ if s is not conjugate to x, which proves (b). Moreover:

$$\psi'(x) = \frac{1}{c \cdot p^a} \sum_{yxy^{-1}=x} \psi(x) = \frac{1}{p^a} \sum_{yxy^{-1}=x} 1 = \frac{\mathrm{Card}(\mathrm{Z}(x))}{p^a}$$

whence $\psi'(x) \not\equiv 0 \pmod{p}$ since $p^a = \mathrm{Card(P)}$ is the largest power of p dividing $\mathrm{Card}(\mathrm{Z}(x))$. □

Lemma 9. *There exists an element ψ of* $\mathrm{A} \otimes \mathrm{V}_p$, *with integer values, such that $\psi(x) \not\equiv 0 \pmod{p}$ for each $x \in \mathrm{G}$.*

Let $(x_i)_{i \in \mathrm{I}}$ be a system of representatives of the p-regular classes (i.e. those consisting of p'-elements). Lemma 8 gives us an element ψ_i of $\mathrm{A} \otimes \mathrm{V}_p$, with integer values, such that

$$\psi_i(x_i) \not\equiv 0 \pmod{p} \qquad \text{and} \qquad \psi_i(x_j) \equiv 0 \pmod{p} \quad \text{for } j \neq i.$$

Put $\psi = \sum \psi_i$. It is clear that ψ belongs to $A \otimes V_p$ and has integer values. For $x \in G$, the p'-component of x is conjugate to a unique x_i. From lemma 7 we obtain

$$\psi(x) \equiv \psi(x_i) \equiv \psi_i(x_i) \not\equiv 0 \,(\text{mod. } p). \qquad \square$$

EXERCISES

10.3. Extend lemma 6 to class functions with values in the ideal gA of A.

10.4. Let $\mathfrak{p}$ be a prime ideal of A such that $\mathfrak{p} \cap \mathbf{Z} = p\mathbf{Z}$ (which is equivalent to saying that $A/\mathfrak{p}$ is a finite field of characteristic p). Let $\chi \in A \otimes R(G)$, let $x \in G$, and let x_r be the p'-component of x. Show that $\chi(x) \equiv \chi(x_r)$ $(\text{mod.}\, \mathfrak{p})$ (same proof as for lemma 7) but that we no longer always have $\chi(x) \equiv \chi(x_r)$ $(\text{mod.}\, pA)$.

10.4 Proof of theorems 18 and 18′

Let $g = p^n l$ be the order of G, with $(p, l) = 1$. It suffices to show that l belongs to $A \otimes V_p$, cf. 10.2.

Let ψ be an element of $A \otimes V_p$ satisfying the conditions of lemma 9. The values of ψ are $\not\equiv 0$ (mod. p). Let $N = \varphi(p^n)$ be the order of the group $(\mathbf{Z}/p^n\mathbf{Z})^*$, so that $\lambda^N \equiv 1$ (mod. p^n) for each integer λ prime to p. Hence $\psi(x)^N \equiv 1$ (mod. p^n) for all $x \in G$, and the function $l(\psi^N - 1)$ has integer values divisible by $lp^n = g$. By lemma 6, this function is an A-linear combination of characters induced from cyclic subgroups of G. Since each cyclic group is p-elementary, we have $l(\psi^N - 1) \in A \otimes V_p$. But $A \otimes V_p$ is an ideal of $A \otimes R(G)$, whence $l\psi^N \in A \otimes V_p$. Subtracting, we get that l belongs to $A \otimes V_p$, which finishes the proof.

10.5 Brauer's theorem

We will say that a subgroup of G is *elementary* if it is p-elementary for at least one prime number p.

Theorem 19. *Each character of* G *is a linear combination with integer coefficients of characters induced from characters of elementary subgroups.*

Let V_p be the subgroup of R(G) defined in th. 18. It suffices to show that the sum V of the V_p, for p prime, is equal to R(G). Now V contains V_p, so the index of V in R(G) divides that of V_p, hence is prime to p by th. 18. Since this is true for all p, this index is equal to 1, which proves the theorem. $\square$

Theorem 20. *Each character of* G *is a linear combination with integer coefficients of monomial characters.*

(Recall that a character is said to be monomial if it is induced from a character of degree 1 of some subgroup.)

This follows from th. 19 and the fact that each character of an elementary group is monomial, since such a group is nilpotent (cf. 8.5, th. 16). □

Remarks

(1) The linear combinations occuring in th. 19 and 20 may have *positive or negative* coefficients. It is in general impossible to write a given character as a linear combination with positive coefficients (integral or even real) of monomial characters, cf. ex. 10.5, below.

(2) Theorem 20 plays an essential role in many applications of representation theory: to a large extent, it gives a reduction of questions pertaining to an arbitrary character χ to the case where χ has degree 1 (hence comes from a character of a cyclic group). It is by this method, for example, that Brauer proved the Artin L-functions are *meromorphic* in the entire complex plane. We will see other applications later.

EXERCISES

10.5. Let χ be an irreducible character of a group G.

(a) Suppose that χ is a linear combination with positive real coefficients of monomial characters. Show that there exists an integer $m \geqslant 1$ such that $m\chi$ is monomial.

(b) Take for G the alternating group $\mathfrak{A}_5$. The corresponding permutation representation is the direct sum of the unit representation and an irreducible representation of degree 4; take for χ the character of this latter representation. If $m\chi$ were induced by a character of degree 1 of a subgroup H, the order of H would be equal to $15/m$, and m could only take the values 1, 3, 5, 15. Moreover, the restriction of χ to H would have to contain a character of degree 1 of multiplicity m (observe that G has no subgroup of order 15). Conclude that χ cannot be a linear combination with positive real coefficients of monomial characters.

10.6. (Suggested by A. Weil.) We want to prove that each $f \in \mathrm{R}(\mathrm{G})$ such that $f(1) = 0$ is a **Z**-linear combination of elements of the form $\mathrm{Ind}_{\mathrm{E}}^{\mathrm{G}}(\alpha - 1)$, where E is an elementary subgroup of G and α is a character of degree 1.

(a) Let $\mathrm{R}_0'(\mathrm{G})$ be the subgroup of R(G) generated by the $\mathrm{Ind}_{\mathrm{E}}^{\mathrm{G}}(\alpha - 1)$, and let $\mathrm{R}'(\mathrm{G}) = \mathbf{Z} + \mathrm{R}_0'(\mathrm{G})$. Show that, if H is a subgroup of G, $\mathrm{Ind}_{\mathrm{H}}^{\mathrm{G}}$ maps $\mathrm{R}_0'(\mathrm{H})$ into $\mathrm{R}_0'(\mathrm{G})$.

(b) Suppose that H is normal in G and that G/H is abelian. Show that $\mathrm{Ind}_{\mathrm{H}}^{\mathrm{G}}$ maps R′(H) into R′(G). [It is enough to show that $\mathrm{Ind}_{\mathrm{H}}^{\mathrm{G}}(1)$ belongs to R′(G), and this follows from the fact that $\mathrm{Ind}_{\mathrm{H}}^{\mathrm{G}}(1)$ is the sum of (G: H) characters of degree 1 of G whose kernel contains H.]

(c) Suppose G is elementary. Let Y be the set of maximal subgroups of G. Show that if H ∈ Y, then H is normal in G, and G/H has prime order [use the fact that G is nilpotent]. Deduce that R(G) is generated by the characters of degree 1 of G together with the $\mathrm{Ind}_H^G(R(H))$, where H runs over Y [apply th. 16]. Show that R′(G) = R(G) [use induction on the order of G, and use (b) to prove that the $\mathrm{Ind}_H^G(R(H))$ are contained in R′(G)].

(d) Return to the general case and denote by X the set of elementary subgroups of G. By th. 19 we have $1 = \sum_{E \in X} \mathrm{Ind}_E^G(f_E)$, with $f_E \in R(E)$. If $\varphi \in R(G)$ this gives

$$\varphi = \sum_{E \in X} \mathrm{Ind}_E^G(\varphi_E) \quad \text{where } \varphi_E = f_E \cdot \mathrm{Res}_E^G(\varphi).$$

If $\varphi(1) = 0$, we have $\varphi_E \in R'_0(E)$ by (c). Conclude that φ belongs to $R'_0(G)$, whence R′(G) = R(G).

CHAPTER 11

Applications of Brauer's theorem

11.1 Characterization of characters

Let B be a subring of **C** and let G be a finite group.

Theorem 21. *Let* φ *be a class function on* G *such that, for each elementary subgroup* H *of* G, *we have* $\mathrm{Res}_H^G \varphi \in B \otimes R(H)$. *Then* $\varphi \in B \otimes R(G)$.

Let X be the set of all elementary subgroups of G. By th. 19, we can write the constant function 1 in the form

$$1 = \sum_{H \in X} \mathrm{Ind}_H^G f_H, \quad \text{with } f_H \in R(H).$$

Multiplying by φ, this gives

$$\varphi = \sum_{H \in X} \varphi \cdot \mathrm{Ind}_H^G f_H = \sum_{H \in X} \mathrm{Ind}_H^G (f_H \cdot \mathrm{Res}_H^G \varphi).$$

Since f_H belongs to R(H) and $\mathrm{Res}_H^G \varphi$ belongs to $B \otimes R(H)$, their product belongs to $B \otimes R(H)$. It follows that φ belongs to $B \otimes R(G)$. □

A similar argument, using Artin's theorem (ch. 9) gives:

Theorem 21′. *Suppose that* B *contains* **Q**. *If* $\mathrm{Res}_H^G \varphi \in B \otimes R(H)$ *for each cyclic subgroup* H *of* G, *then* $\varphi \in B \otimes R(G)$.

Remark. Theorem 21 can be interpreted as a *coherence* property. Suppose that we are given, for each $H \in X$, an element φ_H of $B \otimes R(H)$, and suppose the following properties are satisfied:

(i) If $H' \subset H$, then $\varphi_{H'} = \mathrm{Res}_{H'}^H (\varphi_H)$.
(ii) If $H' = sHs^{-1}$, with $s \in G$, then $\varphi_{H'}$ is obtained from φ_H by means of the isomorphism $x \mapsto sxs^{-1}$.

Then there exists a unique element φ of $B \otimes R(G)$ such that $\mathrm{Res}_H^G \varphi = \varphi_H$ for all $H \in X$.

Theorem 22. *Let* φ *be a class function on* G *such that, for each elementary subgroup* H *of* G, *and each character* χ *of degree* 1 *of* H, *the number*

$$\langle \chi, \mathrm{Res}_H \varphi \rangle_H = \frac{1}{\mathrm{Card}(H)} \sum_{s \in H} \chi(s^{-1})\varphi(s)$$

belongs to B. *Then* φ *belongs to* $B \otimes R(G)$.

Let H be an elementary subgroup of G. Let

$$\mathrm{Res}_H^G \varphi = \sum_{\omega} c_\omega \omega, \quad \text{where } c_\omega = \langle \omega, \mathrm{Res}_H \varphi \rangle_H,$$

be the decomposition of $\mathrm{Res}_H^G \varphi$ into irreducible characters ω of H. By th. 16, each character ω is induced by a character χ_ω of degree 1 of a subgroup H_ω of H. By Frobenius reciprocity, we have

$$c_\omega = \langle \chi_\omega, \mathrm{Res}_{H_\omega}^G \varphi \rangle_{H_\omega}.$$

Since H_ω is an elementary group, the hypothesis on φ insures that c_ω belongs to B. Consequently, $\mathrm{Res}_H \varphi = \sum c_\omega \omega$ belongs to $B \otimes R(H)$, and the result follows by th. 21. □

Corollary. *In order that* φ *be a virtual character* (i.e., $\varphi \in R(G)$), *it is necessary and sufficient that, whenever* H *is an elementary subgroup and* $\chi\colon H \to \mathbf{C}^*$ *is a homomorphism, then* $\langle \chi, \mathrm{Res}_H \varphi \rangle_H \in \mathbf{Z}$.

This is the special case $B = \mathbf{Z}$.

Let Res denote the homomorphism from R(G) into $\bigoplus_{H \in X} R(H)$ defined by the restriction homomorphisms Res_H^G.

Proposition 29. *The homomorphism* $\mathrm{Res}\colon R(G) \to \bigoplus_{H \in X} R(H)$ *is a split injection.*

(A module homomorphism $f\colon L \to M$ is said to be a *split injection* if there exists $r\colon M \to L$ such that $r \circ f = 1$; this is equivalent to saying that f is injective and $f(L)$ is a direct factor of M.)

It is immediate that Res is an injection. To show that it is split, it suffices to prove that its cokernel is torsion free, since the groups under consideration are finitely generated free **Z**-modules. So we must show that, if $f = (f_H)_{H \in X}$ is an element of $\oplus R(H)$, and there exists a non-zero n such that $nf = \mathrm{Res}\ \varphi$, with $\varphi \in R(G)$, then $f \in \mathrm{Im(Res)}$. But this follows from Th. 21, applied to the function φ/n and the ring **Z**. □

[The argument could also be given in terms of *duality*: since the groups involved are finitely generated free $\mathbf{Z}$-modules, showing that Res is split, is equivalent to showing that *its transpose is surjective*. But its transpose is

$$\text{Ind}: \oplus \text{R(H)} \to \text{R(G)},$$

which is indeed surjective by Brauer's Theorem.]

11.2 A theorem of Frobenius

As in Ch. 10, we denote by A the subring of $\mathbf{C}$ generated by the gth roots of unity, where $g = \text{Card(G)}$.

Let n be an integer $\geqslant 1$, and let (g, n) be the g.c.d. of g and n. If f is a function on G, denote by $\Psi^n f$ the function $x \mapsto f(x^n)$. It is easily checked (cf. ex. 9.3) that the operator Ψ^n maps R(G) into itself. Moreover:

Theorem 23. *If f is a class function on* G *with values in* A, *the function* $(g/(g,n))\Psi^n f$ *belongs to* $\text{A} \otimes \text{R(G)}$.

If c is a conjugacy class of G, denote by f_c the *characteristic function* of c, which takes the value 1 on c and 0 on $\text{G} - c$. The function $\Psi^n f_c$ is given by:

$$\Psi^n f_c(x) = \begin{cases} 1 & \text{if } x^n \in c \\ 0 & \text{otherwise} . \end{cases}$$

Each class function with values in A is a linear combination of the f_c. Theorem 23 is thus equivalent to:

Theorem 23′. *For each conjugacy class c of* G, *the function* $(g/(g,n))\Psi^n f_c$ *belongs to* $\text{A} \otimes \text{R(G)}$.

This can be formulated in still another way:

Theorem 23″. *For each conjugacy class c of* G, *and each character χ of* G, *we have* $1/(g,n) \sum_{x^n \in c} \chi(x) \in \text{A}$.

Taking for χ the unit character, this gives:

Corollary 1. *The number of elements $x \in$ G such that $x^n \in c$ is a multiple of* (g, n).

In particular:

Corollary 2. *If n divides the order of* G, *the number of $x \in$ G such that $x^n = 1$ is a multiple of n.*

(We mention at this point a *conjecture* of Frobenius: If the set G_n of those $s \in \text{G}$ such that $s^n = 1$ has n elements, then G_n *is a subgroup of* G.)

PROOF OF THEOREM 23. (R. Brauer.) In view of th. 21, it suffices to show that the restriction of the function $(g/(g,n))\Psi^n f$ to each elementary subgroup H of G belongs to A ⊗ R(H). Now, if h is the order of H, then $g/(g,n)$ is divisible by $h/(h,n)$. So it suffices to show that

$$\frac{h}{(h,n)}\Psi^n(\mathrm{Res}_H f)$$

belongs to A ⊗ R(H), that is, the proof is reduced to the case of *elementary* groups. Since an elementary group is a product of p-groups, it is enough to treat the case of a p-group. Now, using the fact that an irreducible character of such a group is induced by a character of degree 1, we are led finally to proving the following:

Lemma 10. *Let c be a conjugacy class of a p-group* G, *let χ be a character of degree* 1 *of* G, *and let $a_c = \sum_{x^n \in c} \chi(x)$. Then $a_c \equiv 0$ (mod. (g,n)A).*

First, observe that the sum of the a_c (for χ fixed and c variable) is equal to $\sum_{x\in G}\chi(x)$, i.e., to g if $\chi = 1$ and to 0 otherwise. So

$$\textstyle\sum_c a_c \equiv 0 \text{ (mod. } (g,n)).$$

Therefore it is enough to prove lemma 10 for those classes c which are different from the unit class.

Write n in the form $p^a m$, with $(p,m) = 1$. Let p^b be the order of the elements of c, and let C be the set of $x \in$ G such that $x^n \in c$. Since $x^n = x^{p^a m}$ has order $p^b > 1$, and since G is a p-group, the order of x is p^{a+b}. It follows that, if z is an integer $\equiv 1 \,(\mathrm{mod.}\, p^b)$, then $(x^z)^n = x^n$, whence $x^z \in$ C; moreover, we have equality $x^z = x$ if and only if $z \equiv 1 (\mathrm{mod.}\ p^{a+b})$.In other words, the subgroup Γ of $(\mathbf{Z}/p^{a+b}\mathbf{Z})^*$ consisting of elements congruent to 1 mod. p^b *acts freely* on C. Now the set C is partitioned into orbits under the action of Γ, and it suffices to show that the sum of the $\chi(x)$ over each orbit is divisible by (g,n) in the ring A. Such an orbit consists of elements $x^{1+p^b t}$, with $t \in \mathbf{Z}/p^a\mathbf{Z}$. The sum of the values of χ on this orbit is therefore equal to

$$a_c(x) = \chi(x) \sum_{t \bmod. p^a} z^t, \quad \text{where } z = \chi(x^{p^b}).$$

But $\chi(x)$ is a p^{a+b}-th root of unity, and z is a p^a-th root of unity. Therefore

$$\sum_{t \bmod. p^a} z^t = \begin{cases} p^a & \text{if } z = 1, \\ 0 & \text{if } z \neq 1. \end{cases}$$

Consequently $a_c(x)$ is divisible by p^a, and *a fortiori* by (g,n). □

EXERCISE

11.1. Let f be a class function on G with values in **Q** such that $f(x^m) = f(x)$ for all m prime to g. Show that f belongs to **Q** ⊗ R(G) [use th. 21′ to reduce to the cyclic case]. Conclude from th. 23 that, if in addition f has values in $\mathbf{Z}$, then the function $(g/(g,n))\Psi^n f$ belongs to $R(G)$. Apply this to the characteristic function of the unit class.

11.3 A converse to Brauer's theorem

The letters A and g have the same meaning as in the preceding section.

Lemma 11. *Let p be a prime number. Let x be a p'-element of* G, C *the subgroup generated by x, and* P *a Sylow p-subgroup of the centralizer* $Z(x)$ *of x in* G. *Let* H *be a subgroup of* G *containing no conjugate of* $C \times P$, *let ψ be a class function on* H *with values in* A, *and let* $\psi' = \mathrm{Ind}_H^G \psi$. *Then* $\psi'(x) \equiv 0 \pmod{p\,A}$.

Let $S(x)$ be the set of conjugates of x. Then

$$\psi'(x) = \frac{\mathrm{Card}\, Z(x)}{\mathrm{Card}\, H} \sum_{y \in S(x) \cap H} \psi(y).$$

Let $(Y_i)_{i \in I}$ be the distinct H-conjugacy classes contained in $S(x) \cap H$, and choose an element y_i in each Y_i. The number of conjugates of y_i in H is equal to $\mathrm{Card}\, Y_i$, and also equal to $(H : H \cap Z(y_i))$. Therefore

$$\psi'(x) = \frac{\mathrm{Card}\, Z(x)}{\mathrm{Card}\, H} \sum_{i \in I} \mathrm{Card}\, Y_i \cdot \psi(y_i),$$

$$= \sum_{i \in I} n_i \psi(y_i), \quad \text{with } n_i = \frac{\mathrm{Card}\, Z(y_i)}{\mathrm{Card}(H \cap Z(y_i))}.$$

Suppose we have $n_i \not\equiv 0 \pmod{p}$ for some $i \in I$. Then $\mathrm{Card}\, Z(y_i)$ and $\mathrm{Card}(H \cap Z(y_i))$ are divisible by the same power of p; thus a Sylow p-subgroup P_i of $H \cap Z(y_i)$ is also a Sylow p-subgroup of $Z(y_i)$. If C_i is the cyclic group generated by y_i, then $C_i \times P_i$ is contained in H, and is a p-elementary subgroup associated with y_i in the group G. Since y_i and x are conjugate in G, the group $C_i \times P_i$ is conjugate to $C \times P$. This contradicts the hypothesis on H. Thus $n_i \equiv 0 \pmod{p}$ for all i, whence

$$\psi'(x) \equiv 0 \pmod{p\,A}. \qquad \square$$

Theorem 23‴ (J. Green.) *Let $(H_i)_{i \in I}$ be a family of subgroups of* G *such that* $R(G) = \sum_{i \in I} \mathrm{Ind}_{H_i}^G R(H_i)$. *Then each elementary subgroup of* G *is contained in a conjugate of some* H_i.

Let $C \times P$ be a p-elementary subgroup of G. We can assume that this subgroup is *maximal*, and thus associated with a p'-element x of G. If $C \times P$ were not contained in a conjugate of any H_i, the preceding lemma would show $\chi(x) \equiv 0 \pmod{p\,A}$ for all $\chi \in \sum \mathrm{Ind}_{H_i}^G R(H_i)$, in particular for χ equal to the unit character of G, which is absurd. $\square$

In other words, the family of elementary subgroups is "the smallest" for which Brauer's theorem is true.

11.4 The spectrum of $A \otimes R(G)$

Recall that if C is a commutative ring, then the *spectrum* of C, denoted Spec(C), is the set of *prime ideals* of C, cf. Bourbaki, *Alg. Comm.*, Ch. II.

We want to determine the spectrum of the ring $A \otimes R(G)$. (We could also describe that of $R(G)$, but it would be more complicated.)

Let $Cl(G)$ be the set of conjugacy classes of G. The ring $A^{Cl(G)}$ can be identified with the ring of class functions on G with values in A; if f belongs to this ring, and if c is a conjugacy class, let $f(c)$ denote the value of f on an arbitrary element of c. The injections $A \to A \otimes R(G) \to A^{Cl(G)}$ define maps

$$\operatorname{Spec}(A^{Cl(G)}) \to \operatorname{Spec}(A \otimes R(G)) \to \operatorname{Spec}(A).$$

These maps are *surjective*; this follows, for example, from the fact that $A^{Cl(G)}$ is *integral* over A (and even over $\mathbf{Z}$), cf. Bourbaki, *Alg. Comm.*, Ch. IV, §2.

On the other hand, we know that Spec(A) consists of the ideal 0 and the maximal ideals of A. Moreover, if M is maximal in A, the field A/M is finite; its characteristic is called the *residue characteristic* of M.

The spectrum of $A^{Cl(G)}$ can be identified with $Cl(G) \times \operatorname{Spec}(A)$: with each $c \in Cl(G)$ and each $M \in \operatorname{Spec}(A)$ we associate the prime ideal M_c consisting of those $f \in A^{Cl(G)}$ such that $f(c) \in M$. The image of M_c in $\operatorname{Spec}(A \otimes R(G))$ is the prime ideal $P_{M,c} = M_c \cap (A \otimes R(G))$.

Proposition 30. *If*

(i) *with each class* $c \in Cl(G)$ *we associate* $P_{0,c}$,

(ii) *with each p-regular class c and each maximal ideal* M *of* A *with residual characteristic p we associate* $P_{M,c}$,

then we obtain once and only once each prime ideal of $A \otimes R(G)$.

(A conjugacy class is said to be *p-regular* if it consists of p'-elements, cf. 10.1.)

Since $\operatorname{Spec}(A^{Cl(G)}) \to \operatorname{Spec}(A \otimes R(G))$ is surjective (cf. above), each prime ideal $\mathfrak{p}$ of $A \otimes R(G)$ is of the form $P_{M,c}$; since $\mathfrak{p} \cap A$ is M, we see that $\mathfrak{p}$ determines M, and it remains only to determine which pairs of classes c_1 and c_2 are such that $P_{M,c_1} = P_{M,c_2}$. Thus the proposition follows from:

Proposition 30′.

(i) *If* $M = 0$, $P_{0,c_1} = P_{0,c_2}$ *is equivalent to* $c_1 = c_2$.

(ii) *Suppose that* $M \neq 0$ *with residue characteristic p. Let* c_1' (resp. c_2') *be the class consisting of the p'-components of the elements of* c_1 (resp. c_2). *Then* $P_{M,c_2} = P_{M,c_2}$ *is equivalent to* $c_1' = c_2'$.

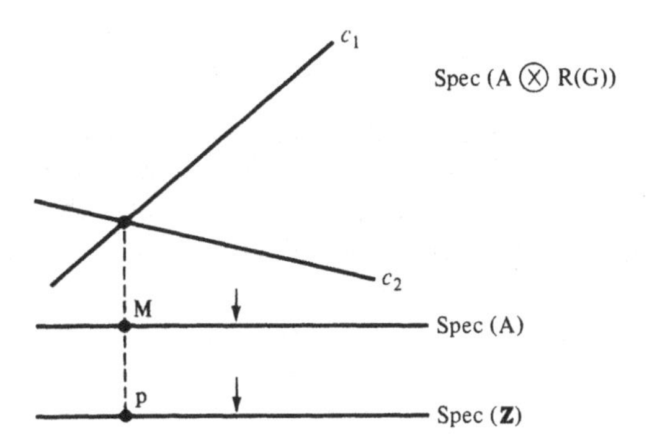

To prove (i) we must show that, if $c_1 \neq c_2$, then there exists an element $f \in \mathrm{A} \otimes \mathrm{R(G)}$ such that $f(c_1) \neq 0$ and $f(c_2) = 0$, and this is clear (take for f the function equal to g on c_1 and 0 elsewhere).

If M has characteristic p, an easy argument, analogous to the proof of lemma 7, shows that $\mathrm{P}_{\mathrm{M},c_1} = \mathrm{P}_{\mathrm{M},c_1'}$ (cf. ex. 10.4). On the other hand, lemma 8 shows that $\mathrm{P}_{\mathrm{M},c_1'} \neq \mathrm{P}_{\mathrm{M},c_2'}$ if $c_1' \neq c_2'$. Whence (ii). □

Remarks

(1) Let I be an ideal of A ⊗ R(G). To show that I is equal to A ⊗ R(G), it suffices to show that I is not contained in any of the prime ideals $\mathrm{P}_{\mathrm{M},c}$; this is the approach taken in the proof of Brauer's theorem (see also ex. 11.7 below).

(2) We can represent Spec(A ⊗ R(G)) graphically as a union of "lines" D_c corresponding to the various classes c, each of these lines representing Spec(A). These lines "intersect" in the following way: D_{c_1} and D_{c_2} have a common point above a maximal ideal M of A with residue characteristic p if and only if the p'-components of c_1 and c_2 are equal.

Proposition 31. Spec(A ⊗ R(G)) *is connected in the Zariski topology.*

(If C is a commutative ring, a subset F of Spec(C) is closed in the Zariski topology if and only if there exists H ⊂ C such that $\mathfrak{p} \in \mathrm{F} \Leftrightarrow \mathfrak{p} \supset \mathrm{H}$.)

Let x be an element of G of order $p_1^{n_1} \cdot p_2^{n_2} \cdots p_k^{n_k}$; x decomposes into a product $x = x_{p_1} \cdot x_{p_2} \cdots x_{p_k}$, where x_{p_i} is of order $p_i^{n_i}$. The classes associated with x and $x_{p_2} \cdots x_{p_k}$ have the same p_1-regular component. Thus, the corresponding "lines" of Spec(A ⊗ R(G)) intersect; moreover, each of these lines is connected, being isomorphic to Spec(A). Proceding step by step until we get to the identity, we see that Spec(A ⊗ R(G)) is connected. □

Corollary. Spec R(G) *is connected.*

Indeed, this is the image of Spec(A ⊗ R(G)) under a continuous map.

EXAMPLE. Take for G the symmetric group $\mathfrak{S}_3$. There are three classes: 1, c_2 (consisting of the elements of order 2), and c_3 (the elements of order 3). There is a unique prime ideal $\mathfrak{p}_2$ in A of residual characteristic 2, and the same holds for 3. The spectrum of A ⊗ R(G) consists of three "lines" which intersect as indicated below:

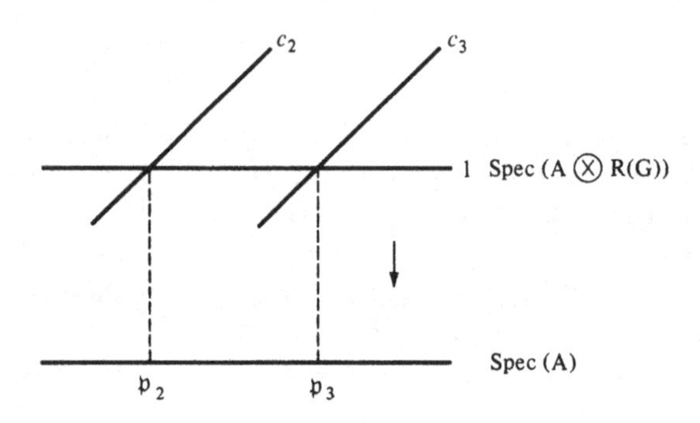

Remark. The results of this section have been extended to compact Lie groups by G. Segal (*Publ. Math. I.H.E.S.*, 34, 1968).

EXERCISES

11.2. Show that the residue field of $P_{M,c}$ is A/M.

11.3. If B is an A-algebra, determine Spec(B ⊗ R(G)) in terms of Spec(B) (use the proof of prop. 30 and 30′).

11.4. Let K be the quotient field of A and let Γ be the Galois group of K/**Q**. We know that Γ is isomorphic to $(\mathbf{Z}/g\mathbf{Z})^*$. Let Γ act on A ⊗ R(G) *via* its action on A, and determine its corresponding action on Spec(A ⊗ R(G)). Obtain Spec(R(G)) by observing that R(G) is the subring of A ⊗ R(G) consisting of those elements fixed by Γ.

11.5. Determine Spec (A[G]) when G is *abelian* (observe that A[G] can be identified with A ⊗ R($\hat{G}$), where $\hat{G}$ is the dual of G, cf. ex. 3.3).

11.6. Let B be the subring of $A^{Cl(G)}$ consisting of those functions f such that, for every maximal ideal M of A with residue characteristic p, and every class c with p-regular component c', we have $f(c) \equiv f(c')(\text{mod}\ .M)$. Show that A ⊗ R(G) ⊂ B, and that these two rings have the same spectrum; give an example where they are distinct.

11.7. Let H be a subgroup of G, and let I_H be the ideal of $A \otimes R(G)$ which is the image of $A \otimes \mathrm{Ind}_H^G$.

(a) Let c be a class of G. Show that I_H is contained in $P_{0,c}$ if and only if $H \cap c = \varnothing$.

(b) Let c be a p-regular class, and let M be a maximal ideal of A containing p. Show that I_H is contained in $P_{M,c}$ if and only if H contains no p-elementary subgroup associated with an element of c.

(c) Obtain from (b) another proof of th. 18 and 23.

CHAPTER 12

Rationality questions

So far we have only studied representations defined over the field $\mathbf{C}$ of complex numbers. In fact, all the proofs of the preceding sections still hold over an *algebraically closed* field of characteristic zero, for example, an algebraic closure of $\mathbf{Q}$. Now we are going to see what happens for fields which are not algebraically closed.

12.1 The rings $R_K(G)$ and $\overline{R}_K(G)$

In this section, K denotes a field of characteristic zero and C an algebraic closure of K. If V is a K-vector space, we let V_C denote the C-vector space $C \otimes_K V$ obtained from V by extending scalars from K to C. If G is a finite group, each linear representation $\rho\colon G \to \mathbf{GL}(V)$ over the field K defines a representation

$$\rho_C\colon G \to \mathbf{GL}(V) \to \mathbf{GL}(V_C)$$

over the field C. In terms of "modules" (cf. 6.1), we have

$$V_C = C[G] \otimes_{K[G]} V.$$

The character $\chi_\rho = \mathrm{Tr}(\rho)$ of ρ is the same as for ρ_C; it is a class function on G with values in K.

We denote by $R_K(G)$ the group generated by the characters of the representations of G over K; it is a subring of the ring $R(G) = R_C(G)$ studied in Ch. 9, 10, 11.

We could also define $R_K(G)$ as the *Grothendieck group* for the category of K[G]-modules of finite type, cf. Part III, Ch. 14.

Proposition 32. *Let* (V_i, ρ_i) *be the distinct (up to isomorphism) irreducible linear representations of* G *over* K, *and let* χ_i *be the corresponding characters. Then*

(a) *The* χ_i *form a basis of* $R_K(G)$.
(b) *The* χ_i *are mutually orthogonal.*

[As usual, this concerns orthogonality with respect to the bilinear form $\langle \varphi, \chi \rangle = (1/g) \sum_{s \in G} \varphi(s^{-1})\chi(s)$.]

It is clear that the χ_i generate $R_K(G)$. On the other hand, if $i \neq j$ we have $\mathrm{Hom}^G(V_i, V_j) = 0$. But in general, if V and W have characters χ_V and χ_W, we have

$$\dim_K \mathrm{Hom}^G(V, W) = \dim_C \mathrm{Hom}^G(V_C, W_C) = \langle \chi_V, \chi_W \rangle,$$

cf. 7.2, lemma 2. It follows that $\langle \chi_i, \chi_j \rangle = 0$ if $i \neq j$, and that $\langle \chi_i, \chi_i \rangle = \dim \mathrm{End}^G(V_i)$ is an integer $\geqslant 1$ (equal to 1 if and only if V_C is irreducible, i.e., if V is *absolutely irreducible*, cf. Bourbaki [8], §13, no. 4). In particular, the χ_i are linearly independent. □

A linear representation of G over C is said to be *realizable over* K (or *rational over* K) if it is isomorphic to a representation of the form ρ_C, where ρ is a linear representation of G over K; this amounts to saying that it can be realized by matrices having coefficients in K.

Proposition 33. *In order that a linear representation of* G *over* C *be realizable over* K, *it is necessary and sufficient that its character belong to* $R_K(G)$.

The condition is obviously necessary. Suppose conversely that it is satisfied, and let χ be the character of the given representation. In view of prop. 32, we have $\chi = \sum n_i \chi_i$, with $n_i \in \mathbf{Z}$, and we obtain:

$$\langle \chi, \chi_i \rangle = n_i \langle \chi_i, \chi_i \rangle \quad \text{for all } i.$$

Since χ is the character of a representation of G over C, the scalar product $\langle \chi, \chi_i \rangle$ is $\geqslant 0$. It follows that n_i is positive, and that the given representation can be realized as the direct sum of the V_i, each repeated n_i times. □

The same argument shows that the realization in question is *unique*, up to K-isomorphism.

In addition to the ring $R_K(G)$, we shall consider the subring $\overline{R}_K(G)$ consisting of those elements of $R(G)$ which have values in K. Obviously, $R_K(G) \subset \overline{R}_K(G)$. Moreover:

Proposition 34. *The group* $R_K(G)$ *has finite index in* $\overline{R}_K(G)$.

First, observe that each irreducible representation of G over C can be realized over a finite extension of K (that generated by the coefficients of a corresponding matrix representation). Hence there exists a finite extension L of K, such that $R_L(G) = R(G)$. Let $d = [L: K]$ be the degree of this extension; the proposition then follows from the following lemma:

Lemma 12. *We have* $d \cdot \overline{R}_K(G) \subset R_K(G)$.

First, let V be a linear representation of G over L with character χ; by restricting scalars we can consider V as a K-vector space (of dimension d times as large) and even as a linear representation of G *over* K. We see immediately that the character of this representation is equal to $\mathrm{Tr}_{L/K}(\chi)$, where $\mathrm{Tr}_{L/K}$ denotes the trace associated with the extension L/K. It follows by linearity that $\mathrm{Tr}_{L/K}(\chi) \in R_K(G)$, for each element χ of $R_L(G)$.

In particular, take $\chi \in \overline{R}_K(G)$, i.e. suppose that the values of χ belong to K. Then $\mathrm{Tr}_{L/K}(\chi) = d \cdot \chi$; hence $d \cdot \chi \in R_K(G)$, and the proof is complete. □

12.2 Schur indices

The results of the preceding section can also be obtained, and even refined, by using the theory of semisimple algebras. We sketch this briefly:

The algebra K [G] is a product of simple algebras A_i, corresponding to the distinct irreducible representations V_i of G over K. If $D_i = \mathrm{Hom}^G(V_i, V_i)$ is the commuting algebra of G in $\mathrm{End}(V_i)$, then D_i is a field (noncommutative, in general), and A_i can be identified with the algebra $\mathrm{End}_{D_i}(V_i)$ of endomorphisms of the D_i-vector space V_i. If $[V_i: D_i] = n_i$, then $A_i \cong M_{n_i}(D_i^\circ)$, where D_i° is the opposite ring of D_i. Moreover, the degree of D_i over its center K_i is a square, say m_i^2; the integer m_i is called the *Schur index* of the representation V_i (or of the component A_i).

Let $s \in G$, and let $\rho_i(s)$ be the corresponding endomorphism of V_i. We have to consider three kinds of "traces" of $\rho_i(s)$:

(a) Its trace as a K-*endomorphism*; this is the element of K denoted above by $\chi_i(s)$;
(b) Its trace as a K_i-*endomorphism*; this is an element of K_i which we will denote by $\varphi_i(s)$;
(c) Its *reduced trace* as an element of the simple algebra A_i (cf. for example [8], no.12.3); this is an element of K_i which we will denote by $\psi_i(s)$.

The various traces are related by the formulas

$$\chi_i(s) = \mathrm{Tr}_{K_i/K}(\varphi_i(s)) \qquad \text{and} \qquad \varphi_i(s) = m_i \psi_i(s).$$

Now let Σ_i be the set of K-homomorphisms of the field K_i into the algebraically closed field C. If $\sigma \in \Sigma_i$, scalar extension from K to C by means of σ makes D_i into a matrix algebra $\mathbf{M}_{m_i}(C)$, and A_i becomes $\mathbf{M}_{n_i m_i}(C)$. Composing $G \to A_i \to \mathbf{M}_{n_i m_i}(C)$, we obtain an irreducible representation of G over C, of degree $n_i m_i$, and with character $\psi_{i,\sigma} = \sigma(\psi_i)$. For fixed i, the characters $\psi_{i,\sigma}$ are *conjugate*: the Galois group of C over K permutes them transitively. Moreover, each irreducible character of G over C is equal to one of the $\psi_{i,\sigma}$. We have

$$\chi_i = \mathrm{Tr}_{K_i/K}(\varphi_i) = \sum_{\sigma \in \Sigma_i} \sigma(\varphi_i) = m_i \sum_{\sigma \in \Sigma_i} \psi_{i,\sigma},$$

which gives the decomposition of χ_i as a sum of irreducible characters over C.

Now let $\chi = \sum_{i,\sigma} d_{i,\sigma} \psi_{i,\sigma}$ be an element of R(G), where the $d_{i,\sigma}$ are integers. In order that χ *have values in* K, it is necessary and sufficient that it be invariant under the Galois group of C over K, i.e., that the $d_{i,\sigma}$ depend only on i. If this is indeed the case, and we let d_i denote their common value, we have

$$\chi = \sum_i d_i \psi_i = \sum_i d_i \chi_i / m_i .$$

Hence we have the following proposition, which refines prop. 34:

Proposition 35. *The characters $\psi_i = \chi_i/m_i$ form a basis of $\overline{R}_K(G)$.*

Let us say that K[G] is *quasisplit* if the D_i are commutative, or, what amounts to the same thing, if the Schur indices m_i are all equal to 1. Then prop. 35 implies:

Corollary. *In order that $R_K(G) = \overline{R}_K(G)$, it is necessary and sufficient that K[G] be quasisplit.*

In particular, we have $R_K(G) = \overline{R}_K(G)$ in each of the following cases:

(i) G is abelian (because then K[G] and the D_i are commutative).
(ii) The Brauer groups of the finite extensions of K are trivial.

EXERCISES

12.1. Show that all the Schur indices for the finite groups considered in Ch. 5 are equal to 1.

12.2. Take for G the alternating group $\mathfrak{A}_4$, cf. 5.7. Show that the decomposition of $\mathbf{Q}[G]$ into simple factors has the form

$$\mathbf{Q}[G] = \mathbf{Q} \times \mathbf{Q}(\omega) \times \mathbf{M}_3(\mathbf{Q}),$$

where $\mathbf{Q}(\omega)$ is the quadratic extension of $\mathbf{Q}$ obtained by adjoining to $\mathbf{Q}$ a cube root of unity ω.

12.3. Take for G the quaternion group $\{\pm 1, \pm i, \pm j, \pm k\}$. The group G has 4 characters of degree 1, with values in $\{\pm 1\}$. On the other hand, the natural embedding of G in the division ring $\mathbf{H_Q}$ of quaternions over $\mathbf{Q}$ defines a surjective homomorphism $\mathbf{Q}[G] \to \mathbf{H_Q}$. Show that the decomposition of $\mathbf{Q}[G]$ into simple components is

$$\mathbf{Q}[G] = \mathbf{Q} \times \mathbf{Q} \times \mathbf{Q} \times \mathbf{Q} \times \mathbf{H_Q}.$$

The Schur index of the last component is equal to 2. The corresponding character ψ is given by

$$\psi(1) = 2, \qquad \psi(-1) = -2, \qquad \psi(s) = 0 \quad \text{for } s \neq \pm 1.$$

Hence K[G] is quasisplit if and only if $K \otimes \mathbf{H_Q}$ is isomorphic to $\mathbf{M}_2(K)$; show that this is equivalent to saying that -1 is a sum of two squares in K.

12.4. Show that the Schur indices m_i divide the index a of the center of G. [Observe that the degree of the irreducible representation with character $\psi_{i,\sigma}$ is $n_i m_i$ and apply prop. 17.] Deduce that $a \cdot \overline{R}_K(G)$ is contained in $R_K(G)$.

12.5. Let L be a finite extension of K. Show that, if L[G] is quasisplit, then [L: K] is divisible by each of the Schur indices m_i.

12.3 Realizability over cyclotomic fields

We keep the notation of the preceding sections, and denote by m the least common multiple of the orders of the elements of G; it is a divisor of g.

Theorem 24 (Brauer). *If* K *contains the* mth *roots of unity, then* $R_K(G) = R(G)$.

In view of prop. 33, this implies:

Corollary. *Each linear representation of G can be realized over K.*

(This result had been conjectured by Schur.)

Let $\chi \in R(G)$. By th. 20 of 10.5, we can write χ in the form

$$\chi = \sum n_i \operatorname{Ind}_{H_i}^G(\varphi_i), \quad (n_i \in \mathbf{Z})$$

where the φ_i are characters of degree 1 of subgroups H_i of G. The values of the φ_i are mth roots of unity; they belong to K. Thus $\varphi_i \in R_K(H_i)$. But, if H is a subgroup of G, it is clear that Ind_H^G maps $R_K(H)$ into $R_K(G)$. Therefore $\operatorname{Ind}_{H_i}^G(\varphi_i) \in R_K(G)$ for all i, which proves the theorem. □

Exercise

12.6. Show that the Schur indices of G (over an arbitrary field) divide the Euler function $\varphi(m)$ [use ex. 12.5].

12.4 The rank of $R_K(G)$

We return now to the case of an arbitrary field K of characteristic zero. We shall determine the rank of $R_K(G)$, or equivalently, *the number of irreducible representations of* G *over* K.

Choose an integer m which is a multiple of the orders of the elements of G (for example, their least common multiple or the order g of G), and let L be the field obtained by adjoining to K the mth roots of unity. We know (cf. for example Bourbaki, *Alg.* V, §11) that the extension L/K is Galois and that its Galois group Gal(L/K) is a subgroup of the multiplicative group $(\mathbf{Z}/m\mathbf{Z})^*$ of invertible elements of $\mathbf{Z}/m\mathbf{Z}$. More precisely, if $\sigma \in \mathrm{Gal}(L/K)$, there exists a unique element $t \in (\mathbf{Z}/m\mathbf{Z})^*$ such that

$$\sigma(\omega) = \omega^t \quad \text{if } \omega^m = 1.$$

We denote by Γ_K the image of Gal(L/K) in $(\mathbf{Z}/m\mathbf{Z})^*$, and if $t \in \Gamma_K$, we let σ_t denote the corresponding element of Gal(L/K). The case considered in the preceding section was that where $\Gamma_K = \{1\}$.

Let $s \in G$, and let n be an integer. Then the element s^n of G depends only on the class of n modulo the order of s, and so *a fortiori* modulo m; in particular s^t is defined for each $t \in \Gamma_K$. The group Γ_K acts as a *permutation group* on the underlying set of G. We will say that two elements s, s' of G are Γ_K*-conjugate* if there exists $t \in \Gamma_K$ such that s' and s^t are conjugate by an element of G. The relation thus defined is an equivalence relation and does not depend upon the choice of m; its classes are called the Γ_K*-classes* (or the K*-classes*) *of* G.

Theorem 25. *In order that a class function f on* G, *with values in* L, *belong to* $K \otimes_{\mathbf{Z}} R(G)$, *it is necessary and sufficient that*

$$(*) \qquad \sigma_t(f(s)) = f(s^t) \quad \textit{for all } s \in G \textit{ and all } t \in \Gamma_K.$$

(In other words, we must have $\sigma_t(f) = \Psi^t(f)$ for all $t \in \Gamma_K$, cf. 11.2.)

Let ρ be a representation of G with character χ. For $s \in G$, the eigenvalues ω_i of $\rho(s)$ are mth roots of unity, and the eigenvalues of $\rho(s^t)$ are the ω_i^t. Thus we have

$$\sigma_t(\chi(s)) = \sigma(\textstyle\sum \omega_i) = \sum \omega_i^t = \chi(s^t),$$

which shows that χ satisfies the condition $(*)$. By linearity, the same is true for all the elements of $K \otimes R(G)$.

Conversely, suppose f is a class function on G satisfying condition $(*)$. Then

$$f = \sum c_\chi \chi, \quad \text{with } c_\chi = \langle f, \chi \rangle,$$

where χ runs over the set of irreducible characters of G. We have to show that the c_χ belong to K, which, according to Galois theory, is equivalent to showing that they are invariant under the σ_t, $t \in \Gamma_K$. But, if φ and χ are two class functions on G, then we have

$$\langle \Psi^t \varphi, \Psi^t \chi \rangle = \langle \varphi, \chi \rangle,$$

as can be easily verified. Whence

$$c_\chi = \langle f, \chi \rangle = \langle \Psi^t f, \Psi^t \chi \rangle = \langle \sigma_t(f), \sigma_t(\chi) \rangle = \sigma_t(\langle f, \chi \rangle) = \sigma_t(c_\chi),$$

which finishes the proof. □

Corollary 1. *In order that a class function f on* G *with values in* K *belong to* $\mathrm{K} \otimes \mathrm{R_K(G)}$, *it is necessary and sufficient that it be constant on the* Γ_K-*classes of* G.

If $f \in \mathrm{K} \otimes \mathrm{R_K(G)}$, then $f(s) \in \mathrm{K}$ for all $s \in \mathrm{G}$, and formula (∗) shows that $f(s) = f(s^t)$ for all $t \in \Gamma_K$. Hence f is constant on the Γ_K-classes of G.

Conversely, suppose that f has values in K, and is constant on the Γ_K-classes of G. Then condition (∗) is satisfied, and we can write

$$f = \sum \langle f, \chi \rangle \chi, \quad \text{with } \langle f, \chi \rangle \in \mathrm{K}$$

as above. Moreover, the fact that f is invariant under the σ_t, $t \in \Gamma_K$, shows that $\langle f, \chi \rangle = \langle f, \sigma_t(\chi) \rangle$, so the coefficients of the two conjugate characters χ and $\sigma_t(\chi)$ are the same. Collecting characters in the same conjugacy class, we can write f as a linear combination of characters of the form $\mathrm{Tr_{L/K}}(\chi)$. Since the latter belong to $\mathrm{R_K(G)}$, cf. 12.1, this proves the corollary.

[*Alternately*: Let Γ_K act on $\mathrm{K} \otimes \mathrm{R(G)}$ by $f \mapsto \sigma_t(f) = \Psi^t(f)$, and observe that the set of fixed points is $\mathrm{K} \otimes \mathrm{R_K(G)}$.] □

Corollary 2. *Let* χ_i *be the characters of the distinct irreducible representations of* G *over* K. *Then the* χ_i *form a basis for the space of functions on* G *which are constant on* Γ_K-*classes, and their number is equal to the number of* Γ_K-*classes.*

This follows from cor. 1. □

Remark. In cor. 1, we can replace $\mathrm{R_K(G)}$ by $\overline{\mathrm{R}}_\mathrm{K}(\mathrm{G})$. Indeed prop. 34 shows that

$$\mathbf{Q} \otimes \mathrm{R_K(G)} = \mathbf{Q} \otimes \overline{\mathrm{R}}_\mathrm{K}(\mathrm{G}), \quad \text{whence } \mathrm{K} \otimes \mathrm{R_K(G)} = \mathrm{K} \otimes \overline{\mathrm{R}}_\mathrm{K}(\mathrm{G}).$$

12.5 Generalization of Artin's theorem

If H is a subgroup of G, it is clear that

$$\mathrm{Res_H}\colon \mathrm{R(G)} \to \mathrm{R(H)} \qquad \text{and} \qquad \mathrm{Ind_H}\colon \mathrm{R(H)} \to \mathrm{R(G)}$$

map $R_K(G)$ into $R_K(H)$ and $R_K(H)$ into $R_K(G)$. So we can ask if the theorems of Artin and Brauer remain valid when R is replaced by R_K. In the case of Artin's theorem, the answer is affirmative:

Theorem 26. *Let* T *be the set of cyclic subgroups of* G. *Then the map*

$$\mathbf{Q} \otimes \mathrm{Ind}: \bigoplus_{H \in T} \mathbf{Q} \otimes R_K(H) \to \mathbf{Q} \otimes R_K(G)$$

defined by the maps $\mathbf{Q} \otimes \mathrm{Ind}_H^G$, $H \in T$, *is surjective.*

The two proofs given in Ch. 9 apply without change. The first is a *duality argument*; one must show that the mapping

$$\mathbf{Q} \otimes \mathrm{Res}: \mathbf{Q} \otimes R_K(G) \to \bigoplus_{H \in T} \mathbf{Q} \otimes R_K(H)$$

is injective, which is clear.

The second proof consists of using the formula

$$g = \sum_{H \in T} \mathrm{Ind}_H^G(\theta_H), \qquad \text{cf. prop. 27 (9.4)},$$

and proving that θ_H belongs to $R_K(H)$. The latter can be verified either by induction on the order of H, or by observing that θ_H has integer values and thus belongs to $\overline{R}_K(H)$; since H is abelian we have $R_K(H) = \overline{R}_K(H)$. Now the identity above shows that the constant function 1 belongs to the image of $\mathbf{Q} \otimes \mathrm{Ind}$. Since this image is an ideal, it must be the whole ring $\mathbf{Q} \otimes R_K(G)$. □

12.6 Generalization of Brauer's theorem

We keep the notation of the preceding sections. It is easy to see that, if X is the family of elementary subgroups of G, the map

$$\mathrm{Ind}: \oplus R_K(H) \to R_K(G)$$

is not, in general, surjective (example: $G = \mathfrak{S}_3$, $K = \mathbf{R}$). It is necessary to replace X by a slightly larger family X_K, that of "Γ_K - elementary" subgroups:

Let p be a prime number. A subgroup H of G is said to be *Γ_K-p-elementary* if it is the semidirect product of a p-group P and a cyclic group C of order prime to p such that*:

($*_K$) *For each* $y \in P$, *there exists* $t \in \Gamma_K$ *such that* $yxy^{-1} = x^t$ *for each* $x \in C$.

* The subgroup C should not be confused with the algebraic closure of K chosen in 12.1; the latter will not appear in this section.

(When $\Gamma_K = \{1\}$, this condition just means that C and P commute, so that $H = C \times P$ is a p-elementary group.) A subgroup of G is said to be Γ_K-*elementary* if it is Γ_K-p-elementary for at least one prime number p.

Denote by X_K (resp. $X_K(p)$) the family of Γ_K-elementary (resp. Γ_K-p-elementary) subgroups of G. Then we have the following analogue of th. 19:

Theorem 27. *The map* Ind: $\bigoplus_{H \in X_K} R_K(H) \to R_K(G)$ *is surjective.*

As in 10.5, we obtain theorem 27 from a more precise result, relative to a fixed prime number p:

Theorem 28. *Let $g = p^n l$ be the order of* G, *where $(p, l) = 1$. The constant function l belongs to the image $V_{K,p}$ of the map*

$$\text{Ind}: \bigoplus_{H \in X_K(p)} R_K(H) \to R_K(G).$$

In particular, the index of $V_{K,p}$ in $R_K(G)$ is finite and prime to p.

The proof of this theorem is completely analogous to that of th. 18′ (to which it reduces when K is algebraically closed). We will give the proof in the next section and, for the time being, just indicate two consequences:

Proposition 36. *Let φ be a class function on* G. *In order that φ belong to $R_K(G)$, it is necessary and sufficient that, for each Γ_K-elementary subgroup* H *of* G, *we have $\text{Res}_H^G \varphi \in R_K(H)$.*

Using th. 27, we have an identity

$$1 = \sum_{H \in X_K} \text{Ind}_H^G f_H, \quad \text{with } f_H \in R_K(H).$$

Multiplying by φ, this gives

$$\varphi = \sum_{H \in X_K} \varphi \cdot \text{Ind}_H^G f_H = \sum_{H \in X_K} \text{Ind}_H^G (f_H \cdot \text{Res}_H^G \varphi).$$

So, if $\text{Res}_H \varphi \in R_K(H)$ for all $H \in X_K$, we have $\varphi \in R_K(G)$; the converse is clear. □

Proposition 37. *If each of the algebras* K[H], $H \in X_K$, *is quasisplit* (cf. 12.2), *the same is true of* K[G].

Let $\varphi \in \overline{R}_K(G)$. For $H \in X_K$, we have $\text{Res}_H^G \varphi \in \overline{R}_K(H)$, and $\overline{R}_K(H)$ is equal to $R_K(H)$ since K[H] is quasisplit (cf. cor. to prop. 35). The preceding proposition then shows that φ belongs to $R_K(G)$. Whence $\overline{R}_K(G) = R_K(G)$, and K[G] is quasisplit. □

EXERCISE

12.7. Show that the map Ind: $\bigoplus_{H \in X_K} \overline{R}_K(H) \to \overline{R}_K(G)$ is surjective. [Use the proof of prop. 36.]

12.7 Proof of theorem 28

We denote by A the subring of L generated by the mth roots of unity.

Lemma 13. *If l belongs to $A \otimes V_{K,p}$, then $l \in V_{K,p}$.*

This is proved by the same argument as the one used in 10.2 for lemma 5. □

Lemma 14. *There are finitely many prime ideals $\mathfrak{p}_1, \ldots, \mathfrak{p}_h$ of A containing p. The quotients $A/\mathfrak{p}_i$ are finite fields of characteristic p, and there exists an integer N such that $pA \supset (\mathfrak{p}_1 \cap \cdots \cap \mathfrak{p}_h)^N$.*

The $\mathfrak{p}_i$ correspond to prime ideals of A/pA, which is a finite ring of characteristic p. The first two assertions follow from this. The third follows from the fact that $(\mathfrak{p}_1 \cap \cdots \cap \mathfrak{p}_h)/pA$ is the *radical* of the artinian ring A/pA, thus is nilpotent. □

Lemma 15. *Let f be a function on G, constant on Γ_K-classes, and with values in gA. Then f can be written in the form*

$$f = \sum \mathrm{Ind}_C^G(\varphi_C), \quad \text{with } \varphi_C \in A \otimes R_K(C),$$

where C runs over the set of cyclic subgroups of G.

Let $\varphi = f/g$. In the notation of lemma 6, we have

$$f = \sum \mathrm{Ind}_C^G(\theta_C \cdot \mathrm{Res}_C^G \varphi),$$

and it remains only to show that $\varphi_C = \theta_C \cdot \mathrm{Res}_C^G \varphi$ belongs to $A \otimes R_K(C)$ for all C. But the values of φ_C are divisible by the order of C; it follows that, if χ is a character of degree 1 of C, we have $\langle \varphi_C, \chi \rangle \in A$. Moreover, the fact that f is constant on Γ_K-classes implies that

$$\langle \varphi_C, \chi \rangle = \langle \Psi^t \varphi_C, \Psi^t \chi \rangle = \langle \varphi_C, \Psi^t \chi \rangle, \quad \text{if } t \in \Gamma_K.$$

The coefficients in φ_C of characters conjugate over K are thus equal, and we can express φ_C as an A-linear combination of traces over K of characters χ; thus $\varphi_C \in A \otimes R_K(C)$. □

Lemma 16. *Let $x, y \in G$ be elements whose p'-components are Γ_K-conjugate. If $f \in A \otimes R_K(G)$, then*

$$f(x) \equiv f(y) \pmod{\mathfrak{p}_i} \qquad \text{for } i = 1, \ldots, h.$$

We know that f is constant on Γ_K-classes (cor. 1 of th. 25). So we can assume that x is the p'-component of y, in which case the same argument applies as in the proof of lemma 7. □

Lemma 17. *Let x be a p'-element of G, let C be the cyclic subgroup generated by x, let $N(x)$ be the set of $y \in G$ such that there exists $t \in \Gamma_K$ with $yxy^{-1} = x^t$, and let P be a Sylow subgroup of $N(x)$. Then:*

(a) $H = C \cdot P$ *is a Γ_K-p-elementary subgroup of* G.
(b) *Each linear representation of* C *over* K *extends to* H.
(c) *The map* Res: $R_K(H) \to R_K(C)$ *is surjective.*

Assertion (a) is clear. To prove (b), it suffices to consider the case of an *irreducible* representation over K. Such a representation can be obtained by choosing a homomorphism $\chi: C \to L^*$, taking as vector space the subfield K_χ generated by $\chi(C)$, and defining $\rho: C \to \mathbf{GL}(K_\chi)$ by the formula

$$\rho(s)\omega = \chi(s)\omega \quad \text{if } s \in C \text{ and } \omega \in K_\chi.$$

The group $\Gamma_K = \mathrm{Gal}(L/K)$ acts K-linearly on K_χ. If $y \in P$, let $t \in \Gamma_K$ be such that $yxy^{-1} = x^t$, and define $\rho(y)$ as the restriction of σ_t to K_χ. One checks that $\rho(y)$ does not depend upon the choice of t, and that

$$\rho(y)\rho(x)\rho(y)^{-1} = \rho(x^t).$$

It follows that the homomorphisms of C and of P into $\mathbf{GL}(K_\chi)$ thus defined extend to a homomorphism of H into $\mathbf{GL}(K_\chi)$, which proves (b). Assertion (c) follows from (b). □

In 10.3 we had $\Gamma_K = \{1\}$, whence $H = C \times P$, so that the lemma above was trivial.

Lemma 18. *Keep the notation of lemma* 17. *Then there exists*

$$\psi \in A \otimes R_K(H)$$

such that the induced function $\psi' = \mathrm{Ind}_H^G \psi$ *has the following properties*:

(i) $\psi'(x) \not\equiv 0 (\mathrm{mod.}\ \mathfrak{p}_i)$ *for* $i = 1, \ldots, h$.
(ii) $\psi'(s) = 0$ *for each p'-element s of* G *which is not Γ_K-conjugate to x.*

Let c be the order of C, and let ψ_C be the function defined on C by $\psi_C(y) = c$ when y has the form x^t, with $t \in \Gamma_K$, and $\psi_C(y) = 0$ otherwise. Then $\psi_C \in A \otimes R_K(C)$: this follows, say, from lemma 15 applied to C. By lemma 17, there exists $\psi \in A \otimes R_K(H)$ such that $\mathrm{Res}_C^H \psi = \psi_C$. We show that ψ works:

If s is a p'-element of G, and if $y \in G$, ysy^{-1} is a p'-element. If $ysy^{-1} \in H$, then $ysy^{-1} \in C$ and $\psi(ysy^{-1})$ is zero whenever ysy^{-1} is not of the form x^t, for $t \in \Gamma_K$. It follows that $\psi'(s) = 0$ if s is not Γ_K-conjugate to

x, which proves (ii). For the rest, let Z be the set of x^t, $t \in \Gamma_K$. Then

$$\psi'(x) = \frac{1}{\text{Card}(H)} \sum_{yxy^{-1} \in Z} c = \frac{\text{Card}(N(x))}{\text{Card}(P)},$$

and since P is a Sylow p-subgroup of $N(x)$ (cf. lemma 17), we see that $\psi'(x)$ is an integer prime to p, whence (i). □

Lemma 19. *There exists $\varphi \in A \otimes V_{K,p}$ such that $\varphi(x) \not\equiv 0 \pmod{\mathfrak{p}_i}$ for each $x \in G$ and each $i = 1, \ldots, h$.*

Let $(x_\lambda)_{\lambda \in \Lambda}$ be a system of representatives for the p'-regular Γ_K-classes i.e., those consisting of p'-elements. For each $\lambda \in \Lambda$, the preceding lemma allows us to construct $\varphi_\lambda \in A \otimes V_{K,p}$ such that

$$\varphi_\lambda(x_\lambda) \not\equiv 0 \ (\text{mod. } \mathfrak{p}_i) \qquad \text{and} \qquad \varphi_\lambda(x_\lambda) = 0 \quad \text{if } \lambda \neq \mu.$$

Put $\varphi = \sum_\lambda \varphi_\lambda$. Then φ belongs to $A \otimes V_{K,p}$ and we have $\varphi(x) \not\equiv 0$ (mod. $\mathfrak{p}_i$) for each p'-element x in G. Lemma 16 shows that the same holds for each x in G. □

Completion of the proof of theorem 28

Let $\varphi \in A \otimes V_{K,p}$ satisfy the conditions of lemma 19. For each $x \in G$ and each i, the class of $\varphi(x)$ mod. $\mathfrak{p}_i$ belongs to the multiplicative group of the field $A/\mathfrak{p}_i$. Since the field $A/\mathfrak{p}_i$ is finite (lemma 14), there is an $M \geqslant 1$ such that $\varphi^M(x) \equiv 1 (\text{mod. } \mathfrak{p}_i)$ for all i and all $x \in G$. Then by lemma 14 we have $\varphi^{MN}(x) \equiv 1$ (mod. pA), and raising φ^{MN} to the power p^n, we obtain $\psi \in A \otimes V_{K,p}$ such that

$$\psi(x) \equiv 1 \ (\text{mod. } p^n A) \quad \text{for all } x \in G.$$

The function $l(\psi - 1)$ thus has values in $p^n l A = gA$. In view of lemma 15, we have $l(\psi - 1) \in A \otimes V_{K,p}$. By subtraction, we obtain that l belongs to $A \otimes V_{K,p}$, and now the theorem follows from lemma 13. □

Exercise

12.8. Determine the spectrum of the ring $A \otimes R_K(G)$. (The result is the same as in 11.4, except that conjugacy classes are replaced by Γ_K-classes.)

CHAPTER 13

Rationality questions: examples

We keep the notation of Ch. 12.

13.1 The field **Q**

Let G be a finite group of order g, and let m be a multiple of the orders of all the elements of G. Take as ground field K the field $\mathbf{Q}$ of rational numbers, and let $\mathbf{Q}(m)$ be the field obtained by adjoining the mth roots of unity to $\mathbf{Q}$. The Galois group of $\mathbf{Q}(m)$ over $\mathbf{Q}$ is the group denoted $\Gamma_{\mathbf{Q}}$ in 12.4; it is a subgroup of the group $(\mathbf{Z}/m\mathbf{Z})^*$. In fact:

Theorem (Gauss–Kronecker). *We have* $\Gamma_{\mathbf{Q}} = (\mathbf{Z}/m\mathbf{Z})^*$.

(This amounts to saying that the mth cyclotomic polynomial Φ_m is *irreducible* over $\mathbf{Q}$.)

We assume this classical result; for a proof, see, for example, Lang [10], p. 204.

Corollary. *Two elements of* G *are* $\Gamma_{\mathbf{Q}}$*-conjugate if and only if the cyclic subgroups they generate are conjugate.*

Applying the results of 12.4, we have:

Theorem 29. *Let f be a class function on* G *with values in* $\mathbf{Q}(m)$.

(a) *In order that f belong to* $\mathbf{Q} \otimes \mathrm{R}(\mathrm{G})$, *it is necessary and sufficient that* $\sigma_t(f) = \Psi^t(f)$ *for each t prime to m.*
(b) *In order that f belong to* $\mathbf{Q} \otimes \mathrm{R}_{\mathbf{Q}}(\mathrm{G})$, *it is necessary and sufficient that f have values in* $\mathbf{Q}$, *and that* $\Psi^t(f) = f$ *for each t prime to m*

(i.e., we must have $f(x) = f(y)$ if x and y generate the same subgroup of G*).*

(Recall that σ_t is the automorphism of $\mathbf{Q}(m)$ which takes an mth root of unity to its tth power, and that $\Psi^t(f)$ is the function $x \mapsto f(x^t)$.)

Corollary 1. *The number of isomorphism classes of irreducible representations of* G *over* **Q** *is equal to the number of conjugacy classes of cyclic subgroups of* G.

This follows from cor. 2 to th. 25.

Corollary 2. *The following properties are equivalent:*

(i) *Each character of* G *has values in* **Q**.
(i′) *Each character of* G *has values in* **Z**.
(ii) *Two elements of* G *which generate the same subgroup are conjugate.*

The equivalence of (i) and (i′) comes from the fact that character values are algebraic integers, thus are elements of **Z** whenever they belong to **Q**. The equivalence of (i) and (ii) follows from th. 29. □

Examples

(1) The symmetric group $\mathfrak{S}_n$ satisfies (ii), hence (i). Moreover, one can show that each representation of $\mathfrak{S}_n$ is *realizable over* **Q**, i.e., that $R(\mathfrak{S}_n) = R_{\mathbf{Q}}(\mathfrak{S}_n)$.

(2) The quaternion group $G = \{\pm 1, \pm i, \pm j, \pm k\}$ satisfies the conditions of the corollary. Hence $\overline{R}_{\mathbf{Q}}(G) = R(G)$; the group $R_{\mathbf{Q}}(G)$ is a subgroup of index 2 of R(G), cf. ex. 12.3.

If H is a subgroup of G, denote by 1_H the unit character of H and by 1_H^G the character of G induced by 1_H (in other words the character of the *permutation representation* on G/H, cf. 3.3, example 2).

Theorem 30. *Each element of* $R_{\mathbf{Q}}(G)$ *is a linear combination, with coefficients in* **Q**, *of characters* 1_C^G, *where* C *runs over the set of cyclic subgroups of* G.

This amounts to saying that $\mathbf{Q} \otimes R_{\mathbf{Q}}(G)$ is generated by the 1_C^G. Since $\mathbf{Q} \otimes R_{\mathbf{Q}}(G)$ is endowed with the nondegenerate bilinear form

$$(\varphi, \psi) \mapsto \langle \varphi, \psi \rangle,$$

we can just as well show that each element θ of $R_{\mathbf{Q}}(G)$ orthogonal to all the 1_C^G is zero. However, we have

$$\langle \theta, 1_C^G \rangle = \langle \mathrm{Res}_C^G \theta, 1_C \rangle = \frac{1}{c} \sum_{s \in C} \theta(s), \quad \text{where } c = \mathrm{Card}\, C.$$

So theorem 30 is equivalent to the following:

Theorem 30′. *If* $\theta \in R_{\mathbf{Q}}(G)$ *is such that* $\sum_{s \in C} \theta(s) = 0$ *for each cyclic subgroup* C *of* G, *then* $\theta = 0$.

We prove this result by induction on Card(G). Let $s \in G$, and let $C(s)$ be the cyclic subgroup of G generated by s. Let $x \in C(s)$; if x generates $C(s)$, we have $\theta(x) = \theta(s)$ since x and s are Γ_Q-conjugate; if x generates a proper subgroup of $C(s)$, the induction hypothesis (applied to the restriction of θ to this subgroup) shows that $\theta(x) = 0$. So we get that

$$\sum_{x \in C(s)} \theta(x) = a \cdot \theta(s),$$

where a is the number of generators of C (s). But by hypothesis we have

$$\sum_{x \in C(s)} \theta(x) = 0,$$

and therefore $\theta(s) = 0$. □

Corollary. *Let* V *and* V′ *be two linear representations of* G *over* **Q**. *In order that* V *be isomorphic to* V′ *it is necessary and sufficient that, for each cyclic subgroup* C *of* G, *we have*

$$\dim V^C = \dim V'^C,$$

where V^C (*resp.* V'^C) *denotes the subspace of* V (*resp.* V′) *consisting of the elements invariant under* C.

The necessity is obvious. To see that the condition is sufficient, let χ and χ' be the characters of V and V′. We have:

$$\dim V^C = \langle \mathrm{Res}^G_C \chi, 1_C \rangle_C$$

and hence $\langle \mathrm{Res}^G_C(\chi - \chi'), 1_C \rangle = 0$ for each C, whence $\chi - \chi' = 0$ by th. 30. Thus V and V′ are isomorphic. □

Remarks

(1) It is not true in general that each element of $R_Q(G)$ is a linear combination *with integer coefficients* of characters 1_H^G, even if H runs over the set of all subgroups of G (cf. ex. 13.4).

(2) Theorem 30 implies the following result: let F/E be a finite Galois extension of number fields, and let χ be the character of a linear representation of Gal(F/E) realizable over **Q**. Then we can write the Artin L-function relative to χ as a product of fractional powers of zeta functions of subfields F_C of F corresponding to cyclic subgroups C of Gal(F/E).

Exercises

13.1. Let G be a cyclic group of order n. For each divisor d of n, denote by G_d the subgroup of G of index d.

(a) Show that G has an irreducible representation over **Q**, unique up to isomorphism, whose kernel is equal to G_d. Let χ_d denote its character; then $\chi_d(1) = \varphi(d)$. The χ_d form an orthogonal basis of $R_Q(G)$.

(b) Define an isomorphism from $\mathbf{Q}[G]$ onto $\prod_{d|n} \mathbf{Q}(d)$.

(c) Put $\psi_d = 1_{G_d}^G$. Show that $\psi_d = \sum_{d'|d} \chi_{d'}$ and that $\chi_d = \sum_{d'|d} \mu(d/d')\psi_{d'}$, where μ denotes the *Möbius function*. Deduce that the ψ_d form a basis for $R_{\mathbf{Q}}(G)$.

13.2. Prove th. 30 by reducing to the cyclic case using th. 26, and then applying ex. 13.1.

13.3. Let ρ be an irreducible representation of G over **Q**, let $A = \mathbf{M}_n(D)$ be the corresponding simple component of $\mathbf{Q}[G]$ (D being a field, not necessarily commutative), and let χ be the character of ρ. Assume that ρ is faithful (i.e., $\ker \rho = 1$) and that every subgroup of G is normal. Let H be a subgroup of G. Show that the permutation reresentation on G/H contains the representation ρ n times if $H = \{1\}$ and 0 times if $H \neq \{1\}$. Conclude that, if $n \geqslant 2$, χ is not contained in the subgroup of $R_{\mathbf{Q}}(G)$ generated by the characters 1_H^G.

13.4. Let E be the quaternion group, C the cyclic group of order 3, and let $G = E \times C$. If $\mathbf{H}_{\mathbf{Q}}$ denotes the usual field of quaternions (over **Q**), show that E and C can be embedded in the multiplicative group $\mathbf{H}_{\mathbf{Q}}^*$. This gives an action of E (resp. C) on the vector space $\mathbf{H}_{\mathbf{Q}}$ by right multiplication (resp. by left multiplication). Obtain from this an irreducible representation ρ of G over **Q** of degree 4. Show that the corresponding simple algebra is isomorphic to $\mathbf{M}_2(K)$, where K is the field of cube roots of unity. Verify the conditions of ex. 13.3 and deduce that the character of ρ is not a linear combination of characters 1_H^G, $H \subset G$.

13.5. Let X and Y be two finite sets on which the group Γ acts. If H is a subgroup of Γ, denote by X^H (resp. Y^H) the set of elements of X (resp. Y) fixed by H. Show that the Γ-sets X and Y are isomorphic if and only if $\mathrm{Card}(X^H) = \mathrm{Card}(Y^H)$ for each subgroup H of Γ. Next, show that the properties listed below are equivalent to each other:

(i) The (linear) permutation representations ρ_X and ρ_Y associated with X and Y are isomorphic.
(ii) For each cyclic subgroup H of Γ, we have $\mathrm{Card}(X^H) = \mathrm{Card}(Y^H)$.
(iii) For each subgroup H of Γ, we have $\mathrm{Card}(X/H) = \mathrm{Card}(Y/H)$.
(iv) For each cyclic subgroup H of Γ, we have $\mathrm{Card}(X/H) = \mathrm{Card}(Y/H)$.

When these properties hold, we shall say that X and Y are *weakly isomorphic*.

[The equivalence of (i) and (ii) is obtained by calculating the characters of ρ_X and ρ_Y. The equivalence of (i) with (iii) and (iv) comes from the fact that $\mathrm{Card}(X/H)$ is the inner product of the character of ρ_X with the character 1_H^Γ.]

Show that, if Γ is cyclic, the Γ-sets X and Y are isomorphic if and only if they are weakly isomorphic. Give an example in the general case of weakly isomorphic sets which are not isomorphic (take for Γ the direct product of two groups of order 2).

13.6. Let X be the set of irreducible characters of G over $\mathbf{Q}(m)$, and let Y be the set of conjugacy classes of G. Let the group $\Gamma_{\mathbf{Q}} = (\mathbf{Z}/m\mathbf{Z})^*$ act on X by $\chi \mapsto \sigma_t(\chi)$ and on Y by $x \mapsto x^t$.

(a) Show that the $\Gamma_{\mathbf{Q}}$-sets X and Y are weakly isomorphic (cf. ex. 13.5).

(b) Show that X (resp. Y) can be identified with the set of homomorphisms from the **Q**-algebra Cent. $\mathbf{Q}[G]$ (resp. $\mathbf{Q} \otimes R(G)$) into $\mathbf{Q}(m)$. Deduce that the $\Gamma_{\mathbf{Q}}$-sets X and Y are isomorphic if and only if the center of $\mathbf{Q}[G]$ is isomorphic to $\mathbf{Q} \otimes R(G)$.

(c) Show that the center of $\mathbf{Q}[G]$ is isomorphic to $\mathbf{Q} \otimes R(G)$ in each of the following cases:

(c_1) G is abelian (use an isomorphism from G onto its dual $\hat{G}$, and observe that $\mathbf{Q}[G] = R(\hat{G})$).

(c_2) G is a p-group and $p \neq 2$ (use the fact that $\Gamma_{\mathbf{Q}}$ is cyclic).
(For an example of a group G such that X and Y are not $\Gamma_{\mathbf{Q}}$-isomorphic, see J. Thompson, *J. of Algebra*, 14, 1970, pp. 1–4.)

13.7. Let p be a prime number $\neq 2$. Let G be a Sylow p-subgroup of $\mathbf{GL}_3(\mathbf{F}_p)$ and let G′ be a nonabelian semidirect product of $\mathbf{Z}/p\mathbf{Z}$ with $\mathbf{Z}/p^2\mathbf{Z}$. Thus $\text{Card}(G) = \text{Card}(G') = p^3$.

(a) Show that G and G′ are not isomorphic.

(b) Construct the irreducible representations of G and G′. Show that $\mathbf{Q}[G]$ and $\mathbf{Q}[G']$ are products of the field **Q**, $p+1$ copies of the field $\mathbf{Q}(p)$, and the matrix algebra $\mathbf{M}_p(\mathbf{Q}(p))$. In particular, $\mathbf{Q}[G]$ and $\mathbf{Q}[G']$ are isomorphic.

(c) Show that $\mathbf{F}_p[G]$ and $\mathbf{F}_p[G']$ are not isomorphic.

13.8. Let $\{C_1, \ldots, C_d\}$ be a system of representatives for the conjugacy classes of cyclic subgroups of G. Show that the characters $1_{C_1}^G, \ldots, 1_{C_d}^G$ form a basis of $\mathbf{Q} \otimes R_{\mathbf{Q}}(G)$.

13.2 The field R

We keep the preceding notation, and take as ground field K the field **R** of real numbers. The corresponding group $\Gamma_{\mathbf{R}}$ is the subgroup $\{\pm 1\}$ of $(\mathbf{Z}/m\mathbf{Z})^*$; two elements x, y of G are $\Gamma_{\mathbf{R}}$-conjugate if and only if y is conjugate to x or to x^{-1}. The automorphism σ_{-1} corresponding to the element -1 of $\Gamma_{\mathbf{R}}$ is just *complex conjugation* $z \mapsto z^*$. If χ is a character of G over C, the general formula $\sigma_t(\chi) = \Psi^t(\chi)$ reduces here to the standard formula

$$\chi(s)^* = \chi(s^{-1}), \quad \text{cf. 2.1, prop. 1}.$$

Theorem 31 (Frobenius-Schur). *Let* $\rho\colon G \to \mathbf{GL}(V)$ *be a linear representation of* G *over* **C** *with character* χ. *In order that* χ *have values in* **R** (*resp. that* ρ *be realizable over* **R**), *it is necessary and sufficient that* V *have a nondegenerate bilinear form* (resp. *symmetric bilinear form*) *invariant under* G.

The group G acts naturally on the dual V′ of V, and it is easy to see that the corresponding character χ' is given by

$$\chi'(s) = \chi(s)^* = \chi(s^{-1}).$$

For χ to have real values, it is necessary and sufficient that $\chi = \chi'$, i.e., that the representations of G in V and V′ are isomorphic. But an isomorphism of V onto V′ corresponds to a nondegenerate bilinear form on V invariant under G. So the existence of such a form is necessary and sufficient for χ to have real values.

Suppose now that ρ is *realizable over* **R**. This is equivalent to saying that we can write V in the form

$$V = V_0 \oplus i\,V_0 = \mathbf{C} \otimes_{\mathbf{R}} V_0,$$

where V_0 is an **R** subspace of V stable under all ρ_s. One knows that there exists a positive definite quadratic form Q_0 on V_0 invariant under G (take the sum of the transforms of an arbitrary positive definite form). By scalar extension, Q_0 defines a quadratic form on V, and the associated bilinear form is nondegenerate, symmetric, and invariant under G.

Conversely, suppose V is endowed with such a form $B(x,y)$. Choose a positive definite hermitian scalar product $(x|y)$ on V, invariant under G; the argument given above shows that such a product exists (cf. 1.3). For each $x \in V$, there exists a unique element $\varphi(x)$ in V such that

$$B(x,y) = (\varphi(x)\,|\,y)^* \quad \text{for all } y \in V.$$

The map $\varphi\colon V \to V$ so defined is antilinear and bijective. Its square φ^2 is an automorphism of V. For $x, y \in V$, we have

$$(\varphi^2(x)\,|\,y) = B(\varphi(x)|\,y)^* = B(y,\varphi(x))^* = (\varphi(y)\,|\varphi(x)).$$

Since $(\varphi(y)\,|\varphi(x)) = (\varphi(x)|\varphi(y))^*$, we get

$$(\varphi^2(x)\,|\,y) = (\varphi^2(y)|\,x)^* = (x|\varphi^2(y)),$$

which means that φ^2 is hermitian. Moreover, the formula

$$(\varphi^2(x)\,|\,x) = (\varphi(x)|\varphi(x))$$

shows that φ^2 is *positive definite*. But we know that, whenever u is hermitian positive definite, there is a unique hermitian positive definite v such that $u = v^2$, and v can be written in the form $P(u)$, where P is a polynomial with real coefficients (if the eigenvalues of u are $\lambda_1,\ldots,\lambda_n$, choose P so that $P(\lambda_i) = \sqrt{\lambda_i}$ for all i). Apply this to $u = \varphi^2$, and put $\sigma = \varphi v^{-1}$. Since $v = P(\varphi^2)$, φ and v commute, and we have $\sigma^2 = \varphi^2 v^{-2} = 1$. Let $V = V_0 \oplus V_1$ be the decomposition of V with respect to the eigenvalues $+1$ and -1 of σ. Since σ is antilinear, multiplication by i maps V_0 onto V_1. Thus

$V = V_0 \oplus i V_0$. On the other hand, the fact that $B(x,y)$ and $(x|y)$ are invariant under G implies that φ, υ, and σ commute with all ρ_s. It follows that V_0 and V_1 are stable under the ρ_s, and we have a realization of V over **R**, which proves th. 31. □

Remarks

(1) Theorem 31 carries over to representations of *compact groups*, cf. Ch. 4. The same is true of the other results in this section.

(2) Denote by $\mathbf{O}_n(\mathbf{C})$ (resp. $\mathbf{O}_n(\mathbf{R})$) the complex (resp. real) orthogonal group in n variables. The last part of the above proof shows, in fact, that each finite (or even compact) subgroup of $\mathbf{O}_n(\mathbf{C})$ is conjugate to one contained in $\mathbf{O}_n(\mathbf{R})$; this is a special case of a general theorem on maximal compact subgroups of Lie groups.

The three types of irreducible representations of G

Let $\rho\colon G \to \mathbf{GL}(V)$ be an irreducible representation of G over **C** of degree n, and let χ be its character. There are three possible cases (mutually exclusive):

(1) One of the values of χ is not real. By restriction of scalars, ρ defines an irreducible representation over **R** of degree $2n$ with character $\chi + \bar{\chi}$. The commuting algebra for this representation is **C**. The corresponding simple component of **R**[G] is isomorphic to $\mathbf{M}_n(\mathbf{C})$; its Schur index is 1.

(2) All values of χ are real, and ρ is realizable by a representation ρ_0 over **R**. The representation ρ_0 is irreducible (and even absolutely irreducible) with character χ. Its commuting algebra is **R**. The corresponding simple component of **R**[G] is isomorphic to $\mathbf{M}_n(\mathbf{R})$; its Schur index is 1.

(3) All values of χ are real, but ρ is not realizable over **R**. By restriction of scalars, ρ defines an irreducible representation over **R** of degree $2n$ and with character 2χ. Its commuting algebra has degree 4 over **R**; it is isomorphic to the field **H** of quaternions. The corresponding simple component of **R**[G] is isomorphic to $\mathbf{M}_n(\mathbf{H})$; its Schur index is 2.

Moreover, every irreducible representation of G over **R** can be obtained by one of the three procedures above: this can be proved by decomposing **R**[G] as a product of simple components, and observing that such a component is of the form $\mathbf{M}_n(\mathbf{R})$, $\mathbf{M}_n(\mathbf{C})$, or $\mathbf{M}_n(\mathbf{H})$. (The fact that **R**[G] is a group algebra is not important here: the same result holds for any semisimple algebra over **R**.)

The types 1, 2, and 3 can be characterized in various ways:

Proposition 38.

(a) *If* G *does not have a nonzero invariant bilinear form on* V, *then* ρ *is of type* 1.

(b) *If such a form does exist, it is unique up to homothety, is nondegenerate, and is either symmetric or alternating. If it is symmetric,* ρ *is of type* 2, *and if it is alternating,* ρ *is of type* 3.

An invariant bilinear form $B \neq 0$ on V corresponds to a G-homomorphism $b \neq 0$ of V into its dual V′. Since V and V′ are irreducible, b is an isomorphism, and this shows that B is nondegenerate. By th. 31, the existence of B means that ρ is of type 2 or 3. Moreover, Schur's lemma shows that B is unique up to homothety. If we write B in the form $B = B_+ + B_-$, with B_+ symmetric and B_- alternating, then B_+ and B_- are also invariant under G. Since B is unique, we have either $B_- = 0$ (and B is symmetric) or $B_+ = 0$ (and B is alternating). By th. 31, the first case corresponds to type 2; thus the second corresponds to type 3. □

Proposition 39. *In order that* ρ *be of type* 1, 2, *or* 3, *it is necessary and sufficient that the number*

$$\langle 1, \Psi^2(\chi)\rangle = \frac{1}{g}\sum_{s\in G} \chi(s^2), \qquad \textit{where } g = \mathrm{Card}(G),$$

be equal to 0, +1, *or* −1, *respectively.*

Let χ_σ^2(resp. χ_α^2) be the character of the symmetric square (resp. the alternating square) of V. Then

$$\chi_\sigma^2 = \frac{1}{2}(\chi^2 + \Psi^2\chi), \qquad \chi_\alpha^2 = \frac{1}{2}(\chi^2 - \Psi^2\chi),$$

cf. 2.1, Prop. 3. Let a_+ and a_- denote the number of times that the symmetric and alternating squares of ρ contain the unit representation. Then

$$a_+ = \langle 1, \chi_\sigma^2\rangle \qquad \text{and} \qquad a_- = \langle 1, \chi_\alpha^2\rangle.$$

On the other hand, the dual of the symmetric (resp. alternating) square of V can be identified with the space of symmetric (resp. alternating) bilinear forms on V. Since dual representations contain the unit representation the same number of times, we obtain from Prop. 38 that:

$$\begin{aligned} &a_+ = a_- = 0 && \text{in case 1,}\\ &a_+ = 1, \quad a_- = 0 && \text{in case 2,}\\ &a_+ = 0, \quad a_- = 1 && \text{in case 3 .} \end{aligned}$$

Since $\langle 1, \Psi^2(\chi)\rangle = a_+ - a_-$, we indeed get 0, +1, and −1 in the three respective cases. The proposition follows. □

Exercises

13.9. If c is a conjugacy class of G, let c^{-1} denote the class consisting of all x^{-1} for $x \in c$. We say that c is *even* if $c = c^{-1}$.

(a) Show that the number of real-valued irreducible characters of G over **C** is equal to the number of even classes of G.

(b) Show that, if G has odd order, the only even class is that of the identity. Deduce that the only real-valued irreducible character of G is the unit character (Burnside).

13.10. Show that the **R**-algebras (Cent. **R**[G]) and **R** ⊗ R(G) are isomorphic.

13.11. Let X_2 and X_3 denote the sets of irreducible characters which are of type 2 and 3, respectively. Show that the integer

$$\sum_{\chi \in X_2} \chi(1) - \sum_{\chi \in X_3} \chi(1)$$

is equal to the number of elements $s \in G$ such that $s^2 = 1$. (Observe that this integer is equal to $\sum \chi(1)\langle 1, \Psi^2(\chi)\rangle = \langle 1, \Psi^2(r_G)\rangle$, where r_G is the character of the regular representation of G.)

Deduce that, if G *has even order*, at least two irreducible characters are of type 2.

13.12. (Burnside). Suppose G has odd order. Let h be the number of conjugacy classes of G. Show that $g \equiv h$ (mod. 16).

[Use the formula $g = \sum_{i=1}^{h} \chi(1)^2$, and observe that the $\chi_i \neq 1$ are conjugate in pairs (cf. ex. 12.9), and that the $\chi_i(1)$ are odd.]

If each prime factor of g is congruent to 1 (mod. 4), show that $g \equiv h$ (mod. 32) by the same method.

Bibliography: Part II

For the general theory of semisimple algebras, see:

[8] N. Bourbaki. *Algèbre*, Chapter VIII, Hermann, Paris, 1958.

[9] C. Curtis and I. Reiner. *Representation Theory of Finite Groups and Associative Algebras*. Interscience Publishers, New York, 1962.

[10] S. Lang. *Algebra*. Addison-Wesley, New York, 1965.

For induced representations, Brauer's theorem, and rationality questions, see [9], [10], and:

[11] W. Feit. *Characters of Finite Groups*. W. A. Benjamin Publishers, New York, 1967.

[12] R. Brauer and J. Tate. On the characters of finite groups. *Ann. of Math., 62* (1955), p. 1–7.

[13] B. Huppert. *Endliche Gruppen I*, Kap. V. Springer-Verlag, 1967.

For applications to the structure of finite groups, see [11], [13], as well as the classical:

[14] W. Burnside. *Theory of Groups of Finite Order*, 2nd edition. Cambridge, 1911; reprinted by Dover Publishers, 1955.

The characters of finite "algebraic" groups are particularly interesting; see:

[15] A. Borel *et al. Seminar on Algebraic Groups and Related Finite Groups*. Lecture Notes in Mathematics, Vol. 131, Springer-Verlag, 1970.

[16] P. Deligne and G. Lusztig. Representations of reductive groups over finite fields, *Ann. of Math.*, 103 (1976), p. 103–161.

A list of some of the principal problems in the theory of characters can be found in:

[17] R. Brauer. Representations of finite groups. *Lectures on Modern Mathematics*, Vol. I, edited by T. Saaty. John Wiley and Sons, New York, 1963.

III

INTRODUCTION TO BRAUER THEORY

We are concerned here with comparing the representations of a finite group in characteristic p with those in characteristic zero. The results, due essentially to Brauer, can be described most conveniently in terms of "Grothendieck groups"; this approach was introduced by Swan (cf. [21], [22]), who also obtained a number of results not discussed here.

Ch. 14 and 15 are preliminary. Ch. 16 contains the statements of the main theorems; they are proved in Ch. 17. In Ch. 18 we express these results in terms of "modular characters." Ch. 19 contains applications to the Artin representations. Some standard definitions are collected in an appendix: Grothendieck groups, projective modules, etc.

The exposition which follows is just an introduction; in particular, the theory of blocks is not touched upon. The interested reader is referred to Curtis–Reiner [9] and Feit's book [20], as well as to the original papers by Brauer, Feit, Green, Osima, Suzuki, and Thompson.

CHAPTER 14

The groups $R_K(G)$, $R_k(G)$, and $P_k(G)$

Notation

In Part III, G denotes a finite group, and m is the l.c.m. of the orders of the elements of G. A field is said to be sufficiently large (relative to G) if it contains the mth roots of unity (cf. 12.3, th. 24).

All modules considered are assumed to be *finitely generated.*

We denote by K a field complete with respect to a discrete valuation v (cf. Appendix) with valuation ring A, maximal ideal $\mathfrak{m}$ and residue field $k = A/\mathfrak{m}$. We assume that K has characteristic zero and that k has characteristic $p > 0$ (so that "reduction modulo $\mathfrak{m}$" goes from characteristic zero to characteristic p).

14.1 The rings $R_K(G)$ and $R_k(G)$

If L is a field we denote by $R_L(G)$ the Grothendieck group of the category of finitely generated L[G]-modules (cf. Appendix). It is a commutative ring with unit with respect to the external tensor product (relative to L). If E is an L[G]-module, we let [E] denote its image in $R_L(G)$; the set of all [E] is denoted by $R_L^+(G)$.

Let S_L denote the set of isomorphism classes of simple L[G]-modules (i.e., irreducible representations of G over L).

Proposition 40. *The family of all elements* [E], *with* $E \in S_L$, *is a basis for the group* $R_L(G)$.

Let R be the free **Z**-module with basis S_L. The family of the various [E], $E \in S_L$, defines a homomorphism $\alpha\colon R \to R_L(G)$. On the other hand, if F is an L[G]-module, and if $E \in S_L$, let $l_E(F)$ denote the number of times which E appears in a composition series of F; it is clear that l_E is an *additive* function of F. Thus there exists a homomorphism $\beta_E: R_L(G) \to \mathbf{Z}$ such that $\beta_E([F]) = l_E(F)$ for all F. The β_E's define a homomorphism

$$\beta: R_L(G) \to R,$$

and it is immediate that α and β are inverses of one another. The proposition follows. □

More generally, the same argument applies to the category of *modules of finite length* over an arbitrary ring.

Note also that the elements of $R_L^+(G)$ are just the linear combinations with non-negative integer coefficients of elements of the basis $([E])_{E \in S_L}$. The preceding discussion applies in particular to the fields K and k. Since K has characteristic zero, the *character* χ_E of a K[G]-module E is already defined; it is an additive function of E. By linearity, we obtain a linear map $x \mapsto \chi_x$ from $R_K(G)$ into the ring of class functions of G. This map is in fact an *isomorphism* of $R_K(G)$ onto the group of *virtual* characters of G over K, and we often identify the two groups (this explains the notation used in 12.1). We also say that χ_x is the *character* (or the *virtual character*) of an element $x \in R_K(G)$.

We will see in Ch. 18 that there is an analogous result for k, in terms of Brauer's *modular characters.*

Remark. If E and E′ are two K[G]-modules such that $[E] = [E']$ in $R_K(G)$, then E and E′ are isomorphic: this follows from the fact that E and E′ are semisimple. The analogous result *is not true* for k[G]-modules if p divides the order of G, because of the existence of modules which are not semisimple.

14.2 The groups $P_k(G)$ and $P_A(G)$

These are defined as the Grothendieck groups of the category of k[G]-modules (resp. of A[G]-modules) which are *projective* (cf. Appendix). Similar definitions are made for $P_k^+(G)$ and $P_A^+(G)$.

If E (resp. F) is a k[G]-module (resp. a projective k[G]-module), then $E \otimes_k F$ is a projective k[G]-module (it suffices to check this when F is free, in which case it is obvious). We obtain thereby an $R_k(G)$-*module* structure on $P_k(G)$.

14.3 Structure of $P_k(G)$

Since k[G] is artinian, we can speak of the *projective envelope* of a k[G]-module M (cf. Gabriel [23] or Giorgiutti [24]). We recall briefly what this means:

A module homomorphism $f: M' \to M$ is called *essential* if $f(M') = M$ and if $f(M'') \neq M$ for all proper submodules M″ of M′. A *projective envelope* of M is a *projective* module P endowed with an essential homomorphism $f: P \to M$.

Proposition 41.

(a) *Every module* M *has a projective envelope which is unique up to isomorphism.*
(b) *If* P_i *is the projective envelope of* M_i $(i = 1, \ldots, n)$, *the direct sum of the* P_i*'s is a projective envelope for the direct sum of the* M_i*'s.*
(c) *If* P *is a projective module, and if* E *is its largest semisimple quotient module, then* P *is a projective envelope for* E.

We prove (a). Write M in the form L/R, where L is projective and R is a submodule of L (we can take L free, for example). For $N \subset R$, let f_N be the canonical homomorphism of L/N onto $M = L/R$. Now let N be minimal in R such that f_N is essential; such a submodule exists, since f_R is essential, and $k[G]$ is artinian. Put $P = L/N$, and let Q be a submodule of L minimal among those whose projection $Q \to P$ is surjective. Since L is projective, the projection $p: L \to P = L/N$ lifts to $q: L \to Q$, and the minimality of Q shows that $q(L) = Q$. Let N′ be the kernel of q. The projection $f_{N'}: L/N' \to L/R$ factors into $L/N' = Q \to L/N \to L/R$ and the two factors are essential. Since N′ is contained in N, the minimality of N implies that $N' = N$, i.e., that $Q \to P$ is an isomorphism. The module L is thus a direct sum $L = N \oplus Q$, which shows that $P = L/N$ is projective. It is then clear that $P \to M$ is a projective envelope of M.

Let $P' \to M$ be another projective envelope of M. Using the fact that P is projective, we see that there exists $g: P \to P'$ such that the triangle

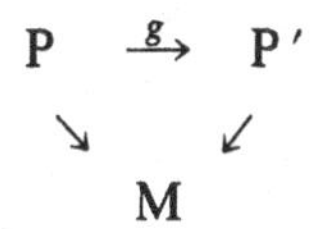

is commutative. The image of $g(P)$ in M is M; since $P' \to M$ is essential, this implies $g(P) = P'$, and so g is surjective. Since P′ is projective, the kernel S of $P \to P'$ is a direct factor in P, which shows that P decomposes into $S \oplus P'$. Using the fact that $P \to M$ is essential, we conclude that $S = 0$, i.e., that $P \to P'$ is an isomorphism. This completes the proof of (a). Assertions (b) and (c) are easy, and left to the reader (see [23], [24] for more details). □

Note that, in case (c), E is the quotient of P by rP, where r is the radical of $k[G]$ (maximum nilpotent ideal); this follows from the fact that the semisimple $k[G]$-modules are those which are annihilated by r. Moreover, by (b), each decomposition of E as a direct sum of simple modules gives a corresponding decomposition of P. Hence we have:

Corollary 1. *Each projective $k[G]$-module is a direct sum of projective indecomposable $k[G]$-modules; this decomposition is unique up to isomorphism. The projective indecomposable $k[G]$-modules are the projective envelopes of the simple $k[G]$-modules.*

Corollary 2. *For each $E \in S_k$, let P_E be a projective envelope of E. Then the $[P_E]$, $E \in S_k$, form a basis of $P_k(G)$.*

Corollary 3. *Two projective $k[G]$-modules P and P′ are isomorphic if and only if their classes [P] and [P′] in $P_k(G)$ are equal.*

More precisely, if $[P] = \sum_{E \in S_k} n_E[P_E]$, the module P is isomorphic to $\prod (P_E)^{n_E}$.

EXERCISE

14.1. Show that $k[G]$ is an *injective* $k[G]$-module. Conclude that a $k[G]$-module is projective if and only if it is injective, and that the projective indecomposable $k[G]$-modules are the injective envelopes of the simple $k[G]$-modules (cf. ex. 14.6).

14.4 Structure of $P_A(G)$

The following result is well known:

Lemma 20. *Let Λ be a commutative ring, and P a $\Lambda[G]$-module. In order that P be projective over $\Lambda[G]$, it is necessary and sufficient that it be projective over Λ, and that there exists a Λ-endomorphism u of P such that*

$$\sum_{s \in G} s \cdot u(s^{-1}x) = x \textit{ for all } x \in P.$$

If P is projective over $\Lambda[G]$ it is projective over Λ: this follows from the fact that $\Lambda[G]$ is Λ-free. Conversely, suppose that the underlying Λ-module P_0 of P is projective, and set $Q = \Lambda[G] \otimes_\Lambda P_0$. The $\Lambda[G]$-module Q is projective. Moreover, the identity map $P_0 \to P$ extends to a surjective $\Lambda[G]$-homomorphism $q\colon Q \to P$. It follows that P is projective if and only if there exists a $\Lambda[G]$-homomorphism $v\colon P \to Q$ such that $q \circ v = 1$. It is easily seen that every $\Lambda[G]$-homomorphism $v\colon P \to Q$ has the form

$$x \mapsto \sum_{s \in G} s \otimes u(s^{-1}x)$$

with $u \in \mathrm{End}_\Lambda(P_0)$. To have $q \circ v = 1$ it is necessary and sufficient to have $\sum_{s \in G} s \cdot u(s^{-1}x) = x$ for all $x \in P$. This proves the lemma. □

Lemma 21. *Suppose that Λ is a local ring, with residue field $k_\Lambda = \Lambda/\mathfrak{m}_\Lambda$.*

(a) *Let* P *be a Λ-free $\Lambda[G]$ module. In order that* P *be $\Lambda[G]$-projective, it is necessary and sufficient that the $k_\Lambda[G]$-module $\bar{P} = P \otimes k_\Lambda$ be projective.*

(b) *Two projective $\Lambda[G]$-modules* P *and* P′ *are isomorphic if and only if the corresponding $k_\Lambda[G]$-modules $\bar{P}$ and $\bar{P}'$ are isomorphic.*

If P is $\Lambda[G]$-projective, then $\bar{P}$ is $k_\Lambda[G]$-projective. Conversely, if this condition is satisfied, the preceding lemma shows that there exists a k_Λ-endomorphism $\bar{u}$ of $\bar{P}$ such that $\sum_{s\in G} s \cdot \bar{u} \cdot s^{-1} = 1$. By lifting $\bar{u}$, we obtain a Λ-endomorphism u of P such that $u' \equiv 1(\text{mod}.\ \mathfrak{m}_\Lambda P)$, where $u' = \sum_{s\in G} s \cdot u \cdot s^{-1}$. Consequently u' is an automorphism of P, which moreover commutes with G. Thus $\sum_{s\in G} s \cdot (uu'^{-1}) \cdot s^{-1} = 1$, which shows that P is projective over $\Lambda[G]$ and proves (a).

If P and P′ are projective, and if $\bar{w}$: $\bar{P} \to \bar{P}'$ is a $k_\Lambda[G]$-homomorphism, the fact that P is projective shows that there exists a $\Lambda[G]$-homomorphism w: P → P′ which lifts $\bar{w}$. If in addition $\bar{w}$ is an isomorphism, then Nakayama's lemma (or an elementary determinant argument) shows that w is an isomorphism. This proves (b). □

We now return to the ring A:

Proposition 42.

(a) *Let* E *be an* A[G]*-module. In order that* E *be a projective* A[G]*-module it is necessary and sufficient that* E *be free over* A *and that the reduction $\bar{E} = E/\mathfrak{m}E$ of* E *be a projective $k[G]$-module.*

(b) *If* F *is a projective $k[G]$-module, there exists a unique* (up to isomorphism) *projective* A[G]*-module whose reduction* mod. $\mathfrak{m}$ *is isomorphic to* F.

Part (a) and the uniqueness in (b) follow from lemmas 20 and 21. It remains to prove *existence* in (b):

Let F be a projective $k[G]$-module. If $n \geqslant 1$ is an integer, let A_n denote the ring $A/\mathfrak{m}^n$; thus $A_1 = k$ and A is the projective limit of the A_n. The rings A_n and $A_n[G]$ are artinian. The arguments in the preceding section show that the $A_n[G]$-module F has a *projective envelope* P_n, and that P_n is free over A_n. The projection $P_n \to F$ factors through $P_n/\mathfrak{m}P_n \to F$, which is surjective. Since F is $k[G]$-projective, there exists a $k[G]$-submodule F′ of $P_n/\mathfrak{m}P_n$ which maps isomorphically onto F. The inverse image P′ of F′ in P_n has image F. Since $P_n \to F$ is essential, it follows that $P' = P_n$, i.e., that $P_n/\mathfrak{m}P_n \to F$ is an *isomorphism*. Moreover, the P_n's form a projective system. Their projective limit P is an A-free A[G]-module, such that $\bar{P} = P/\mathfrak{m}P$ is isomorphic to F. In view of (a), this completes the proof of (b).

Corollary 1. *Every projective* $A[G]$*-module is a direct sum of projective indecomposable* $A[G]$*-modules; this decomposition is unique up to isomorphism. A projective indecomposable* $A[G]$*-module is characterized up to isomorphism by its reduction* mod. $\mathfrak{m}$ *which is a projective indecomposable* $k[G]$*-module* (i.e., the projective envelope of a simple $k[G]$-module).

This follows from the preceding proposition and known results for projective $k[G]$-modules. As a consequence we get:

Corollary 2. *Two projective* $A[G]$*-modules are isomorphic if and only if* $[P] = [Q]$ *in* $P_A(G)$.

Corollary 3. *Reduction* mod. $\mathfrak{m}$ *defines an isomorphism from* $P_A(G)$ *onto* $P_k(G)$*; this isomorphism maps* $P_A^+(G)$ *onto* $P_k^+(G)$.

As a result we may identify $P_A(G)$ and $P_k(G)$.

For a general exposition of projective envelopes in "proartinian" categories, see Demazure-Gabriel [23].

EXERCISES

14.2. Let Λ be a commutative ring, and let P be a $\Lambda[G]$-module which is projective over Λ. Prove the equivalence of the following properties:

(i) P is a projective $\Lambda[G]$-module.

(ii) For each maximal ideal $\mathfrak{p}$ of Λ, the $(\Lambda/\mathfrak{p})[G]$-module $P/\mathfrak{p}P$ is projective.

14.3. (a) Let B be an A-algebra which is free of finite rank over A, and let $\bar{u}$ be an idempotent of $\bar{B} = B/\mathfrak{m}B$. Show the existence of an idempotent of B whose reduction mod. $\mathfrak{m}B$ is equal to $\bar{u}$.

(b) Let P be a projective $A[G]$-module, and let $B = \mathrm{End}^G(P)$. Show that B is A-free, and that $\bar{B}$ can be identified with the algebra of G-endomorphisms of $\bar{P} = P/\mathfrak{m}P$. Deduce from this, and (a), that each decomposition of $\bar{P}$ into a direct sum of $k[G]$-modules lifts to a corresponding decomposition of P.

(c) Use (b) to give another proof of existence in Prop. 42(b). [Write F as a direct factor of a free module $\bar{P}$, lift $\bar{P}$ to a free module, and apply (b).]

14.5 Dualities

Duality between $R_K(G)$ *and* $R_K(G)$

Let E and F be K[G]-modules, and put

$$\langle E, F\rangle = \dim \mathrm{Hom}^G(E, F), \qquad \text{cf. 7.1.}$$

The map $(E, F) \mapsto \langle E, F\rangle$ is "bilinear" (with respect to exact sequences), and so defines a bilinear form

$$R_K(G) \times R_K(G) \to \mathbf{Z},$$

which we denote by $\langle e,f\rangle$ or $\langle e,f\rangle_K$. The classes [E] of simple modules $E \in S_K$ are mutually orthogonal, and $\langle E,E\rangle$ is equal to the dimension d_E of the field $\mathrm{End}^G(E)$ of endomorphisms of E; hence $d_E \geqslant 1$, and equality holds if and only if E is *absolutely simple* (i.e., if the corresponding representation is absolutely irreducible), cf. 12.1.

When K is sufficiently large, it follows from th. 24 that every simple K[G]-module is absolutely simple. Consequently the above bilinear form is *nondegenerate over* **Z**, in the sense that it defines an isomorphism of $R_K(G)$ onto its dual.

Duality between $R_k(G)$ *and* $P_k(G)$

If E is a projective k[G]-module and F an arbitrary k[G]-module, put

$$\langle E, F\rangle = \dim\ \mathrm{Hom}^G(E, F).$$

We thus obtain a bilinear function of E and F (thanks to the assumption that E is projective), hence a bilinear form

$$P_k(G) \times R_k(G) \to \mathbf{Z},$$

denoted $\langle e,f\rangle$ or $\langle e,f\rangle_k$. If $E, E' \in S_k$, we have

$$\mathrm{Hom}^G(P_E, E') = \mathrm{Hom}^G(E, E'),$$

where P_E denotes the projective envelope of E. If $E \neq E'$ we see that $[P_E]$ and $[E']$ are orthogonal; for $E = E'$ we have

$$\langle P_E, E\rangle = \dim\ \mathrm{End}^G(E).$$

As before, $d_E = 1$ if and only if E is absolutely simple.

Suppose that K is sufficiently large, so that k contains the mth roots of unity. We then have $d_E = 1$ for each $E \in S_k$ (see below). Consequently the bilinear form $\langle\ ,\ \rangle_k$ is *nondegenerate over* **Z** and the bases [E] and $[P_E]$ $(E \in S_k)$ are dual to each other with respect to this form.

Remark

The fact that $d_E = 1$ if K is sufficiently large can be proved in various ways:

(1) We can obtain this from th. 24 by "reduction mod. m" once we know that the homomorphism $d\colon R_K(G) \to R_k(G)$ is surjective (cf. Ch. 16, th. 33).

(2) We could also use the fact that Schur indices over k are equal to 1 (cf. 14.6). This reduces the proof to showing that characters of representations of G (over an extension of k) always have *values in* k, and this follows from the fact that they are sums of mth roots of unity.

EXERCISES

14.4. If E is a k[G]-module, we let E′ denote its dual. We define $H^0(G, E)$ as the subspace of E consisting of the elements fixed by G, and $H_0(G, E)$ as the quotient of E by the subspace generated by the $sx - x$, with $x \in E$ and $s \in G$.

(a) Show that, if E is projective, the map $x \mapsto \sum_{s \in G} sx$ defines, by passing to quotients, an isomorphism of $H_0(G, E)$ onto $H^0(G, E)$.

(b) Show that $H^0(G, E)$ is the dual of $H_0(G, E')$. Conclude that $H^0(G, E)$ and $H^0(G, E')$ have the same dimension if E is projective.

14.5. Let E and F be two k[G]-modules, with E projective. Show that

$$\dim \mathrm{Hom}^G(E, F) = \dim \mathrm{Hom}^G(F, E).$$

[Apply part (b) of exercise 14.4 to the projective k[G]-module Hom(E, F), and observe that its dual is isomorphic to Hom(F, E).]

14.6. Let S be a simple k[G]-module and let P_S be its projective envelope. Show that P_S contains a submodule isomorphic to S. [Apply exercise 14.5 with $E = P_S$, $F = S$.] Conclude that P_S is isomorphic to the injective envelope of S, cf. exercise 14.1. In particular, if S is not projective, then S appears *at least twice* in a composition series of P_S.

14.7. Let E be a semisimple k[G]-module, and let P_E be its projective envelope. Show that the projective envelope of the dual of E is isomorphic to the dual of P_E [reduce to the case of a simple module and use exercise 14.6].

14.6 Scalar extensions

If K′ is an extension of K, each K[G]-module E defines by scalar extension a K′[G]-module $K' \otimes_K E$. We thus obtain a homomorphism

$$R_K(G) \to R_{K'}(G).$$

This homomorphism is an injection. This can be seen by determining the image of the canonical basis $\{[E]\}$ $(E \in S_K)$ of $R_K(G)$: if D_E is the (skew) field of endomorphisms of E, the tensor product $K' \otimes D_E$ decomposes as a product of matrix algebras $M_{s_i}(D_i)$, where the D_i are fields. Each of the D_i corresponds to a simple K′[G]-module E'_i, and the image of [E] in $R_{K'}(G)$ is equal to $\sum s_i[E'_i]$. Moreover each simple K′[G]-module is isomorphic to a unique E'_i. This description of $R_K(G) \to R_{K'}(G)$, which generalizes that of 12.2, shows in particular that:

If all the D_E's are *commutative,* the s_i are equal to 1, and the homomorphism $R_K(G) \to R_{K'}(G)$ identifies the first group with a direct factor of the second, i.e., is a *split injection.* If all the $E \in S_K$ are absolutely simple, the $R_K(G) \to R_{K'}(G)$ is an isomorphism.

Analogous results hold for the homomorphisms

$$R_k(G) \to R_{k'}(G), \qquad P_k(G) \to P_{k'}(G)$$

defined by scalar extension from k to k'. The situation is even simpler: the endomorphism field of a simple k[G]-module is always *commutative* and *separable* over k. (This is clear when k is finite, and the general case follows by scalar extension.) Consequently $R_k(G) \to R_{k'}(G)$ is a *split injection.* The same applies for $P_k(G) \to P_{k'}(G)$: since the "scalar extension" functor takes a projective envelope to a projective envelope.

Suppose now that K′ is a *finite* extension of K. Let A′ be the ring of integers of K′ (i.e., the integral closure of A in K′), and k' its residue field. If E is a projective A[G]-module, then $E' = A' \otimes_A E$ is a projective A′[G]-module; moreover, the reduction $k' \otimes_{A'} E'$ of E′ is isomorphic to

$$k' \otimes_A E = k' \otimes_k (k \otimes_A E).$$

The diagram

$$\begin{array}{ccc} P_A(G) & \to & P_{A'}(G) \\ \downarrow & & \downarrow \\ P_k(G) & \to & P_{k'}(G) \end{array}$$

is thus commutative. Since the two vertical arrows are isomorphisms, it follows from the above that the homomorphism $P_A(G) \to P_{A'}(G)$ is a *split injection.*

Remark. The injections $R_K(G) \to R_{K'}(G)$, $R_k(G) \to R_{k'}(G)$, etc., are compatible with the bilinear forms of the preceding section. Moreover, they commute with the homomorphisms *c,d,e* defined in the next chapter.

CHAPTER 15

The *cde* triangle

We shall define homomorphisms *c*, *d*, and *e* which form a commutative triangle:

$$\begin{array}{ccc} P_k(G) & \xrightarrow{c} & R_k(G) \\ & {\scriptstyle e}\searrow \quad \nearrow{\scriptstyle d} & \\ & R_K(G) & \end{array}$$

15.1 Definition of c: $P_k(G) \to R_k(G)$

Associate with each projective $k[G]$-module P the class of P in the group $R_k(G)$. This class is an additive function of P, and so we get a homomorphism

$$c: P_k(G) \to R_k(G)$$

called the *Cartan homomorphism.* If we express c in terms of the canonical bases $[P_S]$ and $[S]$ ($S \in S_k$) of $P_k(G)$ and $R_k(G)$, we obtain a square matrix C, of type $S_k \times S_k$ called the *Cartan matrix* of G (with respect to k). The (S,T) -coefficient C_{ST} of C is the number of times that the simple module S appears in a composition series for the projective envelope P_T of T: we have

$$[P_T] = \sum_{S \in S_k} C_{ST}[S] \quad \text{in } R_k(G).$$

EXERCISE

15.1. Prove that $c(x \cdot y) = x \cdot c(y)$ if $x \in R_k(G)$, $y \in P_k(G)$.

15.2 Definition of d: $R_K(G) \to R_k(G)$

Let E be a K[G]-module. Choose a *lattice* E_1 in E (i.e., a finitely generated A-submodule of E which generates E as a K-module); replacing E_1 by the sum of its images under the elements of G, we can assume that E_1 is stable under G. The reduction $\overline{E}_1 = E_1/\mathfrak{m}E_1$ of E_1 is then a k[G]-module.

Theorem 32. *The image of* $\overline{E}_1$ *in* $R_k(G)$ *is independent of the choice of the stable lattice* E_1.

(Two k[G]-modules $\overline{E}_1$ and $\overline{E}_2$ obtained by reduction of stable lattices E_1 and E_2 need not be isomorphic, cf. ex. 15.1. What the above theorem says is that they have the same *composition factors*.)

Let E_2 be a lattice of E stable under G. We must show that $[\overline{E}_1] = [\overline{E}_2]$ in $R_k(G)$. We begin with a special case:

We have $\mathfrak{m}E_1 \subset E_2 \subset E_1$. Let T be the k[G]-module E_1/E_2. Then we have an exact sequence

$$0 \to T \to \overline{E}_2 \to \overline{E}_1 \to T \to 0,$$

where the homomorphism $T \to \overline{E}_2$ is obtained from multiplication by a generator π of the ideal $\mathfrak{m}$. Passing to $R_k(G)$, we have

$$[T] - [\overline{E}_2] + [\overline{E}_1] - [T] = 0.$$

Thus $[\overline{E}_1] = [\overline{E}_2]$ which proves the theorem in this case.

The general case. Replacing E_2 by a scalar multiple (which does not effect $\overline{E}_2$), we can assume that E_2 is contained in E_1. Thus there exists an integer $n \geqslant 0$ such that

$$\mathfrak{m}^n E_1 \subset E_2 \subset E_1,$$

and we proceed by induction on n. Let $E_3 = \mathfrak{m}^{n-1}E_1 + E_2$. Then

$$\mathfrak{m}^{n-1}E_1 \subset E_3 \subset E_1 \text{ and } \mathfrak{m}E_3 \subset E_2 \subset E_3.$$

By (a) and induction we get

$$[\overline{E}_1] = [\overline{E}_3] = [\overline{E}_2],$$

which proves the theorem. □

It is now clear that the map $E \mapsto [\overline{E}_1]$ extends to a ring homomorphism

$$d\colon R_K(G) \to R_k(G),$$

called the *decomposition* homomorphism. It takes $R_K^+(G)$ into $R_k^+(G)$. The corresponding matrix D (relative to the canonical bases of $R_K(G)$ and $R_k(G)$) is called the *decomposition matrix*. It is a matrix of type $S_k \times S_K$ with nonnegative integer coefficients. For $F \in S_k$ and $E \in S_K$ the corresponding coefficient D_{FE} of D is the number of times that F appears in the reduction mod. $\mathfrak{m}$ of a stable lattice E_1 of E: thus

$$[\bar{E}_1] = \sum_F D_{FE}[F] \qquad \text{in } R_k(G).$$

Remarks

(1) The hypothesis that K be complete plays no role in the proof of th. 32 nor in the definition of the homomorphism *d*.

(2) There are analogous results for *algebraic groups*, cf. *Publ. Sci. I.H.E.S.* no. 34, 1968, pp. 37–52.

Exercises

15.2. Take $p = 2$ and G of order 2. Let $E = K[G]$. Show that E has stable lattices whose reductions are semisimple (isomorphic to $k \oplus k$) and others whose reductions are not semisimple (isomorphic to $k[G]$).

15.3. Let E be a nonzero K[G]-module and E_1 a lattice in E stable under G. Prove the equivalence of the following:

(i) The reduction $\bar{E}_1$ of E_1 is a simple $k[G]$-module.

(ii) Every lattice in E stable under G has the form aE_1 with $a \in K^*$.

Show that these imply that E is a simple K[G]-module.

15.4. (After J. Thompson.) Let E be a **Z**-free **Z**[G]-module, with rank $n \geqslant 2$. Assume that, for each prime number p, the reduction E/pE of E is a simple $(\mathbf{Z}/p\mathbf{Z})[G]$-module.

(a) Show that there is a bilinear form $B(x,y)$ on E with values in **Z** such that $B(x,x) > 0$ for all $x \neq 0$.

(b) Let B be chosen as in (a) and extend it by linearity to the **Q**-vector space $\mathbf{Q} \otimes E$. Show that the set E' of $x \in \mathbf{Q} \otimes E$ such that $B(x,y) \in \mathbf{Z}$ for all $y \in E$ has the form $E' = aE$ with $a \in \mathbf{Q}^*$ (same argument as for ex. 15.3). Conclude that B can be chosen nondegenerate over **Z**, i.e., such that $E' = E$. If $(e_i, \ldots, e_n)$ is a basis of E, the determinant of the matrix of the $B(e_i, e_j)$ is then equal to 1.

(c) Assume that B has been chosen as in (b). Show that there exists $e \in E$ such that $B(x, x) \equiv B(e, x) \pmod{2}$ for all $x \in E$, and that such an e is invariant under G mod. 2E. Conclude that $e \equiv 0 \pmod{2E}$, i.e., that the quadratic form $B(x, x)$ takes only even values.

(d) Obtain from (c) the congruence $n \equiv 0 \pmod{8}$. [Use the fact* that every positive definite integer quadratic form which is even and has discriminant 1 has rank divisible by 8.]

(e) Show that the reflection representation of a Coxeter group of type E_8 has the above properties (cf. Bourbaki, *Gr. et Alg. de Lie*, Ch. VI, §4, no. 10).

* See, for example, *A Course in Arithmetic*, GTM 7, Springer-Verlag (1973), p. 53 and 109.

15.3 Definition of e: $P_k(G) \to R_K(G)$

The functor "tensor product with K" defines a homomorphism from $P_A(G)$ into $R_K(G)$. Combining it with the inverse of the canonical isomorphism $P_A(G) \to P_k(G)$, cf. 14.4, we obtain a homomorphism

$$e: P_k(G) \to R_K(G).$$

Its matrix will be denoted by E; it is of type $S_K \times S_k$.

EXERCISE

15.5. We have $e(d(x) \cdot y) = x \cdot e(y)$ if $x \in R_K(G)$, $y \in P_k(G)$.

15.4 Basic properties of the *cde* triangle

(a) It is *commutative*, i.e., $c = d \circ e$, or equivalently $C = D \cdot E$. This is clear.

(b) The homomorphisms d and e are *adjoints* of one another with respect to the bilinear forms of 14.5:

$$\langle x, d(y)\rangle_k = \langle e(x), y\rangle_K \qquad \text{if } x \in P_k(G), y \in R_K(G).$$

Indeed, we can assume that $x = [\bar{X}]$, where X is a projective A[G]-module, and that $y = [K \otimes_A Y]$, where Y is an A[G]-module which is A-free. The A-module $\mathrm{Hom}^G(X, Y)$ is then free; let r be its rank. Then we have canonical isomorphisms:

$$K \otimes \mathrm{Hom}^G(X, Y) = \mathrm{Hom}^G(K \otimes X, K \otimes Y)$$

and

$$k \otimes \mathrm{Hom}^G(X, Y) = \mathrm{Hom}^G(k \otimes X, k \otimes Y).$$

This shows that $\langle e(x), y\rangle = r = \langle x, d(y)\rangle$.

(c) Assume that K is *sufficiently large*. In view of 14.5, the canonical bases of $P_k(G)$ (resp. of $R_K(G)$) and of $R_k(G)$ (resp. of $R_K(G)$) are dual to each other with respect to the bilinear form $\langle a, b\rangle_k$ (resp. the form $\langle a, b\rangle_K$). This implies that *e can be identified with the transpose of d*; in particular we have $E = {}^tD$. Since $C = D \cdot E = D \cdot {}^tD$, we see that C is a *symmetric matrix*.

EXERCISES

15.6. Let $S, T \in S_k$ and let P_S, P_T be their projective envelopes. We put

$$d_S = \dim\ \mathrm{End}^G(S), \qquad d_T = \dim\ \mathrm{End}^G(T),$$

and let C_{ST} (resp. C_{TS}) be the multiplicity of S (resp. T) in a composition series of P_T (resp. P_S), cf. 15.1.

(a) Show that $C_{ST} d_S = \dim \mathrm{Hom}^G (P_S, P_T)$.

(b) Show that $C_{ST} d_S = C_{TS} d_T$ [apply ex. 14.5]. When K is sufficiently large, the d_S are equal to 1, and we obtain again the fact that the matrix $C = (C_{ST})$ is *symmetric*.

15.7. Keep the notation of Ex. 15.6. Show that either S is projective, $P_S \cong S$ and $C_{SS} = 1$, or $C_{SS} \geqslant 2$ [use ex. 14.6].

15.8. If $x \in P_k(G)$, we have $\langle x, c(x) \rangle_k = \langle e(x), e(x) \rangle_K$. Conclude that, if K is sufficiently large, the quadratic form defined by the Cartan matrix C is positive definite.

15.5 Example: p'-groups

Proposition 43. *Assume that the order of* G *is prime to p. Then*:

(i) *Each* k[G]*-module* (resp. *each* A*-free* A[G]*-module*) *is projective.*

(ii) *The operation of reduction mod.* $\mathfrak{m}$ *defines a bijection from* S_K *onto* S_k.

(iii) *If we identify* S_K *with* S_k *as in* (ii), *the matrices* C, D, E *are all identity matrices.*

(More briefly: the representation theory of the group G is "the same" over k as over K.)

Let E be an A[G]-module which is free over A. We can write E as a quotient L/R of a free A[G]-module L. Since E is A-free, there exists an A-linear projection π of L onto R; since the order g of G is invertible in A, we can replace π by the average $(1/g) \sum_{s \in G} s\pi s^{-1}$ of its conjugates, and the projection thus obtained is A[G]-linear. This shows that E is A[G]-projective. The same argument applies for k[G]-modules. This proves (i), as well as the fact that the Cartan matrix is the identity.

If $E \in S_k$, the projective envelope E_1 of E relative to A[G] is a projective A[G]-module, whose reduction $\overline{E}_1 = E_1/\mathfrak{m}E_1$ is E. If we put $F = K \otimes E_1$, then $d([F]) = [E]$. Since E is simple, this implies that F is simple, thus isomorphic to one of the elements of S_K. We thus obtain a map $E \mapsto F$ of S_k into S_K, and it is clear that this map is the inverse of d. This proves (ii) and (iii). □

Remark. The fact that D is an identity matrix shows that d maps $R_K^+(G)$ onto $R_k^+(G)$; in other words, every linear representation of G over K *can be lifted* to a representation over A, a result which can easily be verified directly (cf. ex. 15.9, below).

EXERCISE

15.9. Suppose that g is prime to p. Let E be a free A-module.

(a) Let $n \geqslant 1$ be an integer, and let

$$\rho_n: G \to \mathbf{GL}(E/\mathfrak{m}^n E)$$

be a homomorphism of G into the group of automorphisms of $E/\mathfrak{m}^n E$. Show that ρ_n can be lifted to

$$\rho_{n+1}: G \to \mathbf{GL}(E/\mathfrak{m}^{n+1} E)$$

and that this lifting is unique, up to conjugation by an automorphism of $E/\mathfrak{m}^{n+1}E$ congruent to 1 mod $.\mathfrak{m}^n$. [Use the fact that the cohomology groups of dimension 1 and 2 of G with values in $\mathrm{End}(E/\mathfrak{m}E)$ are zero.]

(b) Obtain from (a) the fact that every linear representation

$$\rho_1: G \to \mathbf{GL}(E/\mathfrak{m}E)$$

of G over k can be lifted, in an essentially unique way, to a representation of G over A.

15.6 Example: p-groups

Suppose that G is a p-group, of order p^n. We have seen (8.3, cor. to prop. 26) that the only irreducible representation of G in characteristic p is the unit representation. It follows that the artinian ring $k[G]$ is a local ring with residue class field k. The projective envelope of the simple $k[G]$-module k is $k[G]$, i.e., the regular representation of G. The groups $R_k(G)$ and $P_k(G)$ can be both identified with $\mathbf{Z}$, and the Cartan homomorphism $c: \mathbf{Z} \to \mathbf{Z}$ is *multiplication by* p^n. The homomorphism $d: R_K(G) \to \mathbf{Z}$ corresponds to the K-rank; the homomorphism $e: \mathbf{Z} \to R_K(G)$ maps an integer n onto n times the class of the regular representation of G.

15.7 Example: products of p'-groups and p-groups

Suppose that $G = S \times P$, where S has order prime to p, and P is a p-group. We have $k[G] = k[S] \otimes k[P]$. Moreover:

(a) *A* $k[G]$*-module* E *is semisimple if and only if* P *acts trivially on* E.

The sufficiency follows from the fact that every $k[S]$-module is semisimple, cf. 15.5. To prove the necessity, we can assume that E is simple. By 15.6 the subspace E′ of E consisting of elements fixed by P is not zero. Since P is normal in G, the subspace E′ is stable under G, and thus equal to E, which means that P acts trivially.

(b) *A* $k[F]$*-module* E *is projective if and only if it is isomorphic to* $F \otimes k[P]$, *where* F *is a* $k[S]$*-module.*

Since F is a projective $k[S]$-module (cf. 15.5), $F \otimes k[P]$ is a projective $k[G]$-module. Moreover, it is clear that F is the largest quotient of $F \otimes k[P]$

on which P acts trivially. Because of (a) this means that $F \otimes k[P]$ is the projective envelope of F. However, every projective module is the projective envelope of its largest semisimple quotient. We thus see that every projective module has the form $F \otimes k[P]$.

(c) *An* A[G]*-module* $\tilde{E}$ *is projective if and only if it is isomorphic to* $\tilde{F} \otimes A[P]$, *where* $\tilde{F}$ *is an* A*-free* A[S]*-module.*

Clearly a module of the form $\tilde{F} \otimes A[P]$ is projective. The converse is proved by applying (b) to $E = \tilde{E}/\mathfrak{m}\tilde{E}$: if $\tilde{E}$ is projective, we have $E \simeq F \otimes k[P]$, and we know that F can be lifted to an A[S]-module $\tilde{F}$ which is free over A (and even A[S]-projective, cf. above). The module $\tilde{F} \otimes A[P]$ is the projective envelope of $F \otimes k[P]$, and thus is isomorphic to $\tilde{E}$.

Properties (a) and (b) show in particular that the Cartan matrix of G is the *scalar matrix* p^n, where $p^n = \mathrm{Card}(P)$.

CHAPTER 16
Theorems

16.1 Properties of the *cde* triangle

The main result is the following*:

Theorem 33. *The homomorphism* d: $R_K(G) \to R_k(G)$ *is surjective.*

The proof will be given in 17.3.

Remarks

(1) This applies in particular to $k = \mathbf{Z}/p\mathbf{Z}$, taking for K the p-adic field $\mathbf{Q}_p$; the ring A is then the ring $\mathbf{Z}_p$ of p-adic integers.

(2) Roughly speaking, the theorem asserts that every linear representation of G over k can be lifted to characteristic 0 if we are willing to accept "virtual representations", i.e., elements of the Grothendieck group $R_K(G)$. This is an extremely useful result for many applications.

Theorem 34. *The homomorphism* e: $P_k(G) \to R_K(G)$ *is a split injection.*

When K is sufficiently large, e is the transpose of d (cf. 15.4), and the fact that d is surjective implies that e is a split injection. In the general case, let K′ be a finite sufficiently large extension of K, and let k' be its residue field. Consider the diagram:

$$\begin{array}{ccc} P_k(G) & \xrightarrow{e} & R_K(G) \\ \downarrow & & \downarrow \\ P_{k'}(G) & \xrightarrow{e'} & R_{K'}(G). \end{array}$$

* In the first French edition of this book, theorem 33 was stated only for a sufficiently large field K. Claude Chevalley and Andreas Dress have independently observed that it is valid in general.

As we have just seen, e' is a split injection. In view of 14.6, the same is true for $P_k(G) \to P_{k'}(G)$. Their composition is a split injection as well, hence the same holds for e. □

At the same time we have proved:

Corollary 1. *For each finite extension* K′ *of* K, *the homomorphism*

$$P_k(G) \xrightarrow{e} R_K(G) \to R_{K'}(G)$$

is a split injection.

The injectivity of e is equivalent to:

Corollary 2. *Let* P *and* P′ *be projective* A[G]*-modules. If the* K[G]*-modules* $K \otimes P$ *and* $K \otimes P'$ *are isomorphic, then* P *and* P′ *are* A[G]*-isomorphic.*

(Indeed we know that the equality $[P] = [P']$ in $R_A(G) \simeq R_k(G)$ is equivalent to $P \simeq P'$.)

Theorem 35. *Let p^n be the largest power of p dividing the order of* G. *Then every element of $R_k(G)$ divisible by p^n belongs to the image of the Cartan map $c\colon P_k(G) \to R_k(G)$.*

The proof will be given in 17.4.

Corollary 1. *The map $c\colon P_k(G) \to R_k(G)$ is injective, and its cokernel is a finite p-group.*

The second assertion is immediate from th. 35; the first then follows, since $P_k(G)$ and $R_k(G)$ are free $\mathbf{Z}$-modules with the same rank, namely $\mathrm{Card}(S_k)$.

Corollary 2. *If two projective $k(G)$-modules have the same composition factors they are isomorphic.*

This is a restatement of the injectivity of c.

Corollary 3. *Assume* K *is sufficiently large. The Cartan matrix* C *is then symmetric, and the corresponding quadratic form is positive definite. The determinant of* C *is a power of p.*

The quadratic form in question is

$$x \mapsto \langle x, c(x)\rangle_k = \langle x, d(e(x))\rangle_k = \langle e(x), e(x)\rangle_K, \qquad x \in P_k(G).$$

Since the form $\langle a, b\rangle_K$ is clearly positive definite, and e is injective (th. 34), we see that the above form is also positive definite. The determinant of C is thus > 0. This implies that $\det(C)$ is a power of p, since the cokernel of c is a p-group. □

Remarks

(1) The above argument shows that the injectivity of c follows from that of e.

(2) Theorem 35 is equivalent to the assertion that there exists a homomorphism $c'\colon R_k(G) \to P_k(G)$ such that $c \circ c' = p^n$ (which implies $c' \circ c = p^n$).

(3) The exponent n in th. 35 is best possible, cf. ex. 16.3.

EXERCISES

16.1. Show that, when K is not complete, theorem 33 remains valid provided K is sufficiently large. (If $\hat{K}$ denotes the completion of K, observe that the homomorphism $R_K(G) \to R_{\hat{K}}(G)$ is an isomorphism, and apply th. 33 to $\hat{K}$.)

16.2. Show that $d\colon R_{\mathbf{Q}}(G) \to R_{\mathbf{Z}/5\mathbf{Z}}(G)$ is not surjective if G is cyclic of order 4.

16.3. Let H be a Sylow p-subgroup of G. Show that, if E is a projective $k[G]$-module, then E is a free $k[H]$-module (cf. 15.6), and so dim E is divisible by p^n. Conclude that the map $[E] \mapsto \dim E$ defines, by passing to quotients, a surjective homomorphism Coker $(c) \to \mathbf{Z}/p^n\mathbf{Z}$. In particular, the element p^{n-1} of $R_k(G)$ does not belong to the image of c.

16.2 Characterization of the image of e

An element of G is said to be *p-singular* if it is not p-regular (cf. 10.1), i.e., if its order is divisible by p. Recall also that every element of $R_K(G)$ can be identified with a *class function* on G, namely its character (cf. 12.1 and 14.1).

Theorem 36. *The image of* $e\colon P_k(G) \to R_K(G)$ *consists of those elements of* $R_K(G)$ *whose character is zero on the p-singular elements of* G.

We even have the more precise result:

Theorem 37. *Let* K′ *be a finite extension of* K. *In order that an element of* $R_{K'}(G)$ *belong to the image of* $P_A(G) = P_k(G)$ *under e, it is necessary and sufficient that its character take values in* K, *and be zero on the p-singular elements of* G.

For the proof, see 17.5.

EXERCISE

16.4. (Swan.) Let Λ be a Dedekind domain with quotient field F. Assume that, for each prime number p dividing the order of G, there exists a prime ideal $\mathfrak{p}$ of Λ such that $\Lambda/\mathfrak{p}$ has characteristic p. Let P be a projective $\Lambda[G]$-module. Show that $F \otimes P$ is a *free* $F[G]$-module. [Apply th. 36 to the modules obtained from P by completion at such primes $\mathfrak{p}$. Deduce that the character of $F \otimes P$ is zero off the identity element of G.]

This exercise applies in particular to the case where Λ is the ring of integers of an algebraic number field.

16.3 Characterization of projective A[G]-modules by their characters

Such a characterization amounts to determining those representations of G over K which contain a lattice stable under G which is *projective* as an A[G]-module. In other words, it amounts to characterizing the image of $P_K^+(G) = P_A^+(G)$ under e. Only partial results are known. First:

Lemma 22. *Let $x \in P_A(G)$, and let $n \geqslant 1$ be an integer. If $nx \in P_A^+(G)$, we have $x \in P_A^+(G)$.*

This is clear: if $r = \mathrm{Card}(S_k)$, then $P_A(G)$ can be identified with $\mathbf{Z}^r$ and $P_A^+(G)$ with $\mathbf{N}^r$, cf. 14.3 and 14.4. □

Proposition 44. *Let K′ be a finite extension of K, and let A′ be the ring of integers of K′. Assume the following two conditions on an element x of $R_{K'}(G)$:*

(a) *The character of x has values in K.*
(b) *There exists an integer $n \geqslant 1$, such that nx arises, by scalar extension, from a projective A′[G]-module.*

Then x arises from a projective A[G]-module, uniquely determined up to isomorphism.

Let $N = [K'\colon K] = [A'\colon A]$. Let E′ be a projective A′[G]-module with image nx in $R_{K'}(G)$, and let E be the A[G]-module obtained from E′ by restriction to A[G]. One checks easily that the character of $K \otimes E$ is equal to nN times that of x.

Thus

$$e([E]) = nN \cdot x \quad \text{in } R_{K'}(G).$$

By th. 36, the character of $e([E])$ is zero on the p-singular elements of G; hence the same is true for x. So, by th. 37, we have $x = e(y)$, with $y \in P_A(G)$. Since e is injective (th. 34), this implies $[E] = nN \cdot y$, and lemma 22 shows that y belongs to $P_A^+(G)$. Consequently, there exists a projective A[G]-module Y such that $[K \otimes Y] = x$ in $R_K(G)$; the uniqueness of Y (up to isomorphism) follows from cor. 2 to th. 34. □

One can ask whether $e(P_A^+(G)) = e(P_A(G)) \cap R_K^+(G)$. This is not true in general (cf. ex. 16.5 and 16.7). However, we have the following criterion:

Proposition 45. *Suppose the following condition is satisfied:*

(R) *There exists a finite extension* K′ *of* K, *with residue field* k', *such that* $d(\mathrm{R}^+_{\mathrm{K}'}(\mathrm{G})) = \mathrm{R}^+_{k'}(\mathrm{G})$.

Then we have $e(\mathrm{P}^+_{\mathrm{A}}(\mathrm{G})) = e(\mathrm{P}_{\mathrm{A}}(\mathrm{G})) \cap \mathrm{R}^+_{\mathrm{K}}(\mathrm{G})$.

By prop. 44 it is enough to prove that

$$e(\mathrm{P}^+_{\mathrm{A}}(\mathrm{G})) = e(\mathrm{P}_{\mathrm{A}}(\mathrm{G})) \cap \mathrm{R}^+_{\mathrm{K}}(\mathrm{G})$$

when K is sufficiently large, in which case condition (R) just means that d maps $\mathrm{R}^+_{\mathrm{K}}(\mathrm{G})$ *onto* $\mathrm{R}^+_k(\mathrm{G})$. Now let

$$x \in e(\mathrm{P}_{\mathrm{A}}(\mathrm{G})) \cap \mathrm{R}^+_{\mathrm{K}}(\mathrm{G}).$$

Since $x \in e(\mathrm{P}_{\mathrm{A}}(\mathrm{G}))$, we can write x as

$$x = \sum_{\mathrm{E} \in \mathrm{S}_k} n_{\mathrm{E}}\, e([\tilde{\mathrm{P}}_{\mathrm{E}}]),$$

where $\tilde{\mathrm{P}}_{\mathrm{E}}$ denotes a projective A[G]-module whose reduction mod. $\mathfrak{m}$ is the projective envelope P_{E} of E (cf. 14.4). We must show that the integers n_{E} are nonnegative. By (R), for each $\mathrm{E} \in \mathrm{S}_k$ there exists $z_{\mathrm{E}} \in \mathrm{R}^+_{\mathrm{K}}(\mathrm{G})$ such that $d(z_{\mathrm{E}}) = [\mathrm{E}]$. Since $x \in \mathrm{R}^+_k(\mathrm{G})$, we have $\langle x, z_{\mathrm{E}}\rangle_{\mathrm{K}} \geqslant 0$. On the other hand, the fact that d and e are adjoint shows that $\langle x, z_{\mathrm{E}}\rangle_{\mathrm{K}} = n_{\mathrm{E}}$. In particular n_{E} is non-negative, and the proof is complete. □

Combining prop. 45 and th. 36, we get:

Corollary. *Suppose that* G *satisfies condition* (R) *of prop.* 45. *A linear representation of* G *over* K *comes from a projective* A[G]*-module, if and only if its character vanishes on the p-singular elements of* G.

Remark. Condition (R) is equivalent to the following:

(R′) *If* K *is sufficiently large, every simple* k[G]*-module is the reduction mod.* $\mathfrak{m}$ *of a* K[G]*-module* (necessarily simple).

(In other words, each irreducible linear representation of G over k lifts to K.)

Theorem 38. (*Fong-Swan*). *Suppose that* G *is p-solvable, i.e., has a normal composition series whose factors are either p-groups or groups of order prime to p. Then* G *satisfies conditions* (R) *and* (R′) *above.*

For the proof, see 17.6.

EXERCISES

16.5. With notation as in prop. 44, show that

$$\mathrm{P}^+_{\mathrm{A}}(\mathrm{G}) = \mathrm{P}^+_{\mathrm{A}'}(\mathrm{G}) \cap \mathrm{P}_{\mathrm{A}}(\mathrm{G}) = \mathrm{P}^+_{\mathrm{A}'}(\mathrm{G}) \cap \mathrm{R}_{\mathrm{K}}(\mathrm{G}).$$

16.6. Show that, for K sufficiently large, condition (R) is equivalent to the condition $e(\mathrm{P}_{\mathrm{A}}^{+}(\mathrm{G})) = e(\mathrm{P}_{\mathrm{A}}(\mathrm{G})) \cap \mathrm{R}_{\mathrm{K}}^{+}(\mathrm{G})$. (Observe that an element x of $\mathrm{P}_k(\mathrm{G})$ belongs to $\mathrm{P}_k^{+}(\mathrm{G})$ if and only if $\langle x,y\rangle_k \geqslant 0$ for all $y \in \mathrm{R}_k^{+}(\mathrm{G})$.)

16.7. Take for G the group $\mathbf{SL}(\mathrm{V})$ where V is a vector space of dimension 2 over the field $\mathbf{F}_p = \mathbf{Z}/p\mathbf{Z}$. Show that the natural representations of G in the ith symmetric powers V_i of V are absolutely irreducible for $i < p$. (Since the number of p-regular classes of G is p, it follows that these are, up to isomorphism, all the irreducible representations of G, cf. 18.2, cor. 2 to th. 42.) Give examples where these representations cannot be lifted to characteristic zero even over a sufficiently large field K. (For $p = 7$, $i = 4$, we have $\dim \mathrm{V}_i = 5$, and 5 does not divide the order of G; hence V_i cannot be lifted.)

16.4 Examples of projective A[G]-modules: irreducible representations of defect zero

In this section we assume that K is sufficiently large.

Proposition 46. *Let* E *be a simple* K[G]*-module, and let* P *be a lattice in* E *stable under* G. *Assume that the dimension* N *of* E *is divisible by the largest power p^n of p dividing the order g of* G. *Then*:

(a) P *is a projective* A[G]*-module.*
(b) *The canonical map* $\mathrm{A}[\mathrm{G}] \to \mathrm{End}_{\mathrm{A}}(\mathrm{P})$ *is surjective, and its kernel is a direct factor in* A[G] (as a two-sided ideal).
(c) *The reduction* $\overline{\mathrm{P}} = \mathrm{P}/\mathfrak{m}\mathrm{P}$ *of* P *is a simple and projective* A[G]*-module.*

Observe that (a) implies (cf. th. 37):

Corollary. *The character* χ_{E} *of* E *is zero on p-singular elements of* G.

First of all, since N is divisible by p^n, *the quotient* N/g *belongs to the ring* A. This enables us to apply Fourier inversion (6.2., prop. 11) without introducing any "denominators," i.e., within the ring A. More precisely, let s_{P} be the endomorphism of P defined by $s \in \mathrm{G}$; if $\phi \in \mathrm{End}_{\mathrm{A}}(\mathrm{P})$, the *trace* $\mathrm{Tr}(s_{\mathrm{P}}^{-1}\phi)$ of $s_{\mathrm{P}}^{-1}\phi$ belongs to A, so we can define the element

$$u_\phi = \frac{\mathrm{N}}{g} \sum_{s\in \mathrm{G}} \mathrm{Tr}(s_{\mathrm{P}}^{-1}\phi)s \quad \text{of the ring A [G]}.$$

It follows from prop. 11 that u_ϕ has image $1 \otimes \phi$ in $\mathrm{End}_{\mathrm{K}}(\mathrm{E})$, and 0 in $\mathrm{End}_{\mathrm{K}}(\mathrm{E}')$ for each simple K[G]-module E′ not isomorphic to E. In particular, u_ϕ has image ϕ in $\mathrm{End}_{\mathrm{A}}(\mathrm{P})$, which proves (b). Assertion (a) then follows from the elementary fact that P is projective over the ring $\mathrm{End}_{\mathrm{A}}(\mathrm{P})$; the same argument works for (c). □

Remark. In the language of *block theory* (cf. [9], [20]), prop. 46 is the case of a *block with a unique irreducible character* (or of *defect zero*).

EXAMPLE. If G is a semisimple linear group over a finite field of characteristic p, there exists a linear irreducible representation of G (over $\mathbf{Q}$) whose degree is equal to p^n; it is the *special representation* of G discovered by R. Steinberg (cf. *Canad. J. of Math.*, 8, 1956, p. 580–591 and 9, 1957, p. 347–351). By a result of Solomon–Tits it may be realized as the homology representation of top dimension for the *Tits building* associated with G*.

EXERCISES

16.8. Take $G = \mathfrak{A}_4$, cf. 5.7. Show that, for $p = 2$, the group G has no irreducible representation of the type described by prop. 46, but that there is such a representation for $p = 3$. Same question for $\mathfrak{S}_4$.

16.9. Let $S \in S_k$. Prove the equivalence of the following properties:
(i) S is a projective k[G]-module.
(ii) S is isomorphic to the reduction mod. $\mathfrak{m}$ of a module P satisfying the conditions of prop. 46.
(iii) The diagonal coefficient C_{SS} of the Cartan matrix of G is equal to 1. (For the equivalence of (i) and (iii), see ex. 15.7.)

* Cf. L. Solomon, The Steinberg character of a finite group with a BN-pair. *Theory of Finite Groups*, edited by R. Brauer and C.-H Sah. W. A. Benjamin, New York, 1969, p. 213–221.

CHAPTER 17
Proofs

17.1 Change of groups

Let H be a subgroup of G. We have already defined *restriction* and *induction* homomorphisms relative to R_K:

$$\mathrm{Res}_H^G: R_K(G) \to R_K(H) \quad \text{and} \quad \mathrm{Ind}_H^G: R_K(H) \to R_K(G).$$

The same definitions apply to R_k and P_k: by restriction, every $k[G]$-module defines a $k[H]$ module, which is projective if the given module is projective. Passing to Grothendieck groups, we get homomorphisms

$$\mathrm{Res}_H^G: R_k(G) \to R_k(H) \quad \text{and} \quad \mathrm{Res}_H^G: P_k(G) \to P_k(H).$$

On the other hand, if E is a $k[H]$-module, then $\mathrm{Ind}\ E = k[G] \otimes_{k[H]} E$ is a $k[G]$-module (said to be *induced* by E), which is projective if E is projective. Hence we have homomorphisms

$$\mathrm{Ind}_H^G: R_k(H) \to R_k(G) \quad \text{and} \quad \mathrm{Ind}_H^G: P_k(H) \to P_k(G).$$

Using the associativity of the tensor product, we easily obtain the formula

$$(*) \qquad \mathrm{Ind}_H^G(x \cdot \mathrm{Res}_H^G y) = \mathrm{Ind}_H^G(x) \cdot y$$

in each of the following situations:

(a) $x \in R_K(H), y \in R_K(G)$ and $\mathrm{Ind}_H^G(x) \cdot y \in R_K(G)$,

(b) $x \in R_k(H), y \in R_k(G)$ and $\mathrm{Ind}_H^G(x) \cdot y \in R_k(G)$,

(c) $x \in R_k(H), y \in P_k(G)$ and $\mathrm{Ind}_H^G(x) \cdot y \in P_k(G)$.

[Case (c) makes sense because $P_k(G)$ is a *module* over $R_k(G)$.]

Moreover, the homomorphisms c, d, e of Ch. 15 *commute* with the homomorphisms Res_H^G and Ind_H^G.

EXERCISE

17.1. Extend the definitions of Res_H^G and Ind_H^G to the case of a *homomorphism* $H \to G$ whose kernel has order prime to p (cf. ex. 7.1).

17.2 Brauer's theorem in the modular case

Theorem 39. *Let* X *be the set of all* Γ_K*-elementary subgroups of* G (*cf.* 12.6). *The homomorphisms*

$$\mathrm{Ind}\colon \bigoplus_{H \in X} R_k(H) \to R_k(G)$$

and

$$\mathrm{Ind}\colon \bigoplus_{H \in X} P_k(H) \to P_k(G)$$

defined by the Ind_H^G, *for* $H \in X$, *are surjective.*

(In other words, th. 27 holds for R_k and P_k.)

Let 1_K (resp. 1_k) denote the identity element of the ring $R_K(G)$ (resp. $R_k(G)$). We have $d(1_K) = 1_k$. By th. 27 we can write 1_K in the form

$$1_K = \sum_{H \in X} \mathrm{Ind}_H(x_H) \quad \text{with } x_H \in R_K(H).$$

Applying d, and using the fact that d commutes with Ind_H^G, we obtain an analogous formula for 1_k:

$$1_k = \sum_{H \in X} \mathrm{Ind}_H(x'_H), \quad \text{with } x'_H = d(x_H) \in R_k(H).$$

For $y \in R_k(G)$ (resp. $P_k(G)$), we get by multiplication:

$$y = 1_k \cdot y = \sum_{H \in X} \mathrm{Ind}_H(x'_H) \cdot y = \sum_{H \in X} \mathrm{Ind}_H^G(x'_H \cdot \mathrm{Res}_H^G y),$$

which proves the theorem. □

Corollary. *If* K *is sufficiently large, each element of* $R_k(G)$ (resp. *of* $P_k(G)$) *is a sum of elements of the form* $\mathrm{Ind}_H(y_H)$, *where* H *is an elementary subgroup of* G, and y_H belongs to $R_k(H)$ (resp. to $P_k(H)$).

Indeed, when K is sufficiently large, then X is just the set of all elementary subgroups of G.

Remark. The argument used in the proof of th. 39 applies to many other situations (cf. Swan [21], §§ 3,4.) For example it gives the following analogue of Artin's theorem (cf. th. 26):

Theorem 40. *Let* T *be the set of all cyclic subgroups of* G. *The homomorphisms*

$$\mathbf{Q} \otimes \mathrm{Ind}\colon \bigoplus_{\mathrm{H} \in \mathrm{T}} \mathbf{Q} \otimes \mathrm{R}_k(\mathrm{H}) \to \mathbf{Q} \otimes \mathrm{R}_k(\mathrm{G})$$

and

$$\mathbf{Q} \otimes \mathrm{Ind}\colon \bigoplus_{\mathrm{H} \in \mathrm{T}} \mathbf{Q} \otimes \mathrm{P}_k(\mathrm{H}) \to \mathbf{Q} \otimes \mathrm{P}_k(\mathrm{G})$$

are surjective.

17.3 Proof of theorem 33

We have to show that $d\colon \mathrm{R}_\mathrm{K}(\mathrm{G}) \to \mathrm{R}_k(\mathrm{G})$ is surjective. By th. 39, $\mathrm{R}_k(\mathrm{G})$ is generated by the various $\mathrm{Ind}_\mathrm{H}^\mathrm{G}(\mathrm{R}_k(\mathrm{H}))$, where H is Γ_K-elementary. Since d commutes with $\mathrm{Ind}_\mathrm{H}^\mathrm{G}$, it is enough to show that $\mathrm{R}_k(\mathrm{H}) = d(\mathrm{R}_\mathrm{K}(\mathrm{H}))$. Hence we are *reduced to the case where* G *is* Γ_K*-elementary*. In this case we have the following more precise result:

Theorem 41. *Let l be a prime number. Assume that* G *is the semidirect product of an l-group* P *by a cyclic normal subgroup* C *of order prime to l. Then every simple k*[G]*-module* E *can be lifted* (i.e., is the reduction mod $\mathfrak{m}$ of an A-free A [G]-module).

(In other words, d maps $\mathrm{R}_\mathrm{K}^+(\mathrm{G})$ *onto* $\mathrm{R}_k^+(\mathrm{G})$.)

Suppose $l \neq p$. Let C_p be the p-Sylow subgroup of C, and let E′ be the vector subspace of E consisting of those elements fixed by C_p. Since C_p is a p-group, we have $\mathrm{E}' \neq 0$, cf. 8.3., prop. 26. Since C_p is normal in G, the space E′ is stable under G. Thus $\mathrm{E}' = \mathrm{E}$, which means that *C_p acts trivially* on E, and that the representation of G in E comes from a representation of G/C_p. Since the order of G/C_p is prime to p, it is immediate that such a representation can be lifted (cf. 15.5).

Suppose now that $l = p$. We proceed by induction on the order of G. Since C has order prime to p, the representation of C in E is semisimple. Decompose it into a direct sum of isotypic k[C]-modules (cf. 8.1 prop. 24):

$$\mathrm{E} = \bigoplus_\alpha \mathrm{E}_\alpha.$$

The group G permutes the E_α *'s*; since E is simple, G permutes transitively the nonzero E_α *'s*. Let E_β be one of these, and let G_β be the subgroup of G consisting of those elements s such that $s\mathrm{E}_\beta = \mathrm{E}_\beta$. It is clear that E_β is a $k[\mathrm{G}_\beta]$-module and that E is isomorphic to the corresponding induced module $\mathrm{Ind}_{G_\beta}^{G}(E_\beta)$. Moreover, G_β is the semidirect product of a subgroup of P and the group C. If $E_\beta \neq E$, we have $G_\beta \neq G$, and the induction hypothesis applied to G_β shows that E_β can be lifted; the same is then true for E.

Thus we may assume that E is an *isotypic* $k[C]$*-module*. Let ρ denote the homomorphism from $k[G]$ into $\mathrm{End}_k(E)$ which defines the $k[G]$-module structure on E. The fact that E is $k[C]$-isotypic is equivalent to saying that the image of $k[C]$ under ρ is a *field* k', which is a finite extension of k. The restriction of ρ to C is a homomorphism $\phi\colon C \to k'^*$, and k' is generated over k by ϕ (C). The module E is thus endowed with the structure of a k'-vector space. Now choose an element $v \neq 0$ of E invariant under P; again this is possible since P is a p-group, cf. 8.3, prop. 26. For $x \in C$, $s \in P$, put ${}^s x = sxs^{-1}$. We have

$$\rho(s)(\phi(x) \cdot v) = \rho(sxs^{-1})\rho(s) \cdot v = \phi({}^s x) \cdot v.$$

Hence the subspace $k'v$ of E generated by the $\phi(x)\cdot v, x \in C$, is stable under C and P, thus is equal to E. Hence $\dim_{k'} E = 1$. This allows us to identify E with k' in such a way that v becomes the unit element of k'. For all $t \in G$ $\rho(t)$ is an endomorphism σ_t of the k-vector space k'. For $s \in P$ we have $\sigma_s(1) = 1$ by construction. Moreover, the above formula shows that

$$\sigma_s(\phi(x)) = \phi({}^s x) \quad \text{for all } x \in C,$$

hence

$$\sigma_s(\phi(x)\phi(x')) = \sigma_s(\phi(x))\sigma_s(\phi(x')) \quad \text{for all } x, x' \in C.$$

Since k' is generated by the $\phi(x)$, we get

$$\sigma_s(aa') = \sigma_s(a)\sigma_s(a') \quad \text{if } a, a' \in k';$$

in other words, σ_s is an *automorphism* of the field k' and the map $s \mapsto \sigma_s$ is a *homomorphism* $\sigma\colon P \to \mathrm{Gal}(k'/k)$, where the latter denotes the Galois group of k'/k. The *lifting* of E is now easy to define: let K′ be the unramified extension of K corresponding to the residue extension k'/k, and let A′ be the ring of integers of K′. The canonical isomorphism

$$\mathrm{Gal}(K'/K) \xrightarrow{\sim} \mathrm{Gal}(k'/k)$$

gives an action of P on K′ and on A′ (using σ). On the other hand, the homomorphism $\phi\colon C \to k'^*$ lifts uniquely (using, say, multiplicative representatives) to a homomorphism $\tilde{\phi}\colon C \to A'^*$, which gives an action of C on A′ by multiplication. It is then immediate (from uniqueness) that we still have

$$\sigma_s(\tilde{\phi}(x)) = \tilde{\phi}({}^s x) \quad \text{for } x \in C, s \in P.$$

This means that the actions of C and P on A′ combine to give an action of G. Endowed with such an A[G]-module structure, A′ is the desired lifting. □

Remark. When K is sufficiently large, we only need th. 41 in the case where G is *elementary*, thus a direct product of C with P. The above proof becomes much simpler: the group P acts trivially on the simple module E, which can thus be viewed as a simple k[C]-module and lifted without difficulty.

17.4 Proof of theorem 35

Let p^n be the largest power of p dividing the order of G. We have to show that the cokernel of $c\colon P_k(G) \to R_k(G)$ is killed by p^n. We distinguish two cases:

(a) K *is sufficiently large*

By the cor. to th. 39, $R_k(G)$ is generated by the $\mathrm{Ind}_H^G(R_k(H))$ with H elementary. We are thus reduced to the case where G is *elementary*, hence decomposes as a product $S \times P$, where S has order prime to p and P is a p-group. We have seen in 15.7 that the Cartan matrix of such a group is the scalar matrix p^n. The theorem follows in this case.

(b) *General case*

Let K′ be a finite sufficiently large extension of K, with residue field k'. Scalar extension from k to k' gives us a commutative diagram:

$$\begin{array}{ccccccc} 0 \to & P_k(G) & \to & P_{k'}(G) & \to & P & \to 0 \\ & \downarrow c & & \downarrow c' & & \downarrow \gamma & \\ 0 \to & P_k(G) & \to & R_{k'}(G) & \to & R & \to 0, \end{array}$$

where $P = P_{k'}(G)/P_k(G)$ and $R = R_{k'}(G)/R_k(G)$. Whence the exact sequence:

$$0 \to \mathrm{Ker}(c) \to \mathrm{Ker}(c') \to \mathrm{Ker}(\gamma) \to \mathrm{Coker}(c) \to \mathrm{Coker}(c').$$

By (a), Coker(c') is killed by p^n. Since $P_{k'}(G)$ and $R_{k'}(G)$ have the same rank, it follows that c' is injective, whence the same is true for c, and so Coker(c) is finite. But we know (cf. 14.6) that $P_k(G) \to P_{k'}(G)$ is a split injection. The group P is thus **Z**-free, and so is Ker(γ). Since Ker$(c') = 0$, and Coker(c) is finite, the exact sequence above shows that Ker$(\gamma) = 0$; hence Coker(c) embeds in Coker(c'). Since the latter is killed by p^n, the same is true of Coker(c), which proves the theorem. □

17.5 Proof of theorem 37

By extending K′ if necessary, we can assume that K′ is *sufficiently large*.

(i) *Necessity*

Let E be a projective A[G]-module, and let χ be the character of the K′[G]-module $K' \otimes_{A'} E$. If $s \in G$ is p-singular, *we must show that* $\chi(s) = 0$. Replacing G by the cyclic subgroup generated by s, we can assume G is cyclic, hence of the form $S \times P$, where S has order prime to p, and P is a p-group. By 15.7, E is isomorphic to $F \otimes A'[P]$, where F is an A′-free A′[S]-module. The character χ of $K' \otimes E$ is thus equal to $\psi \otimes r_P$, where ψ is a character of S and r_P is the character of the regular representation of P. Such a character is evidently zero off S, so in particular $\chi(s) = 0$.

(ii) *Sufficiency (first part)*

Let $y \in R_{k'}(G)$, let χ be the corresponding virtual character, and suppose $\chi(s) = 0$ for every p-singular element s of G.

We will show that y *belongs to* $P_{A'}(G)$ (where this group is identified with a subgroup of $R_{k'}(G)$ by means of e).

By the cor. to th. 39, we have

$$1 = \sum \operatorname{Ind}(x_H), \qquad \text{with } x_H \in R_{K'}(H),$$

where H runs over the set of all elementary subgroups of G. Multiplying by y, we get:

$$y = \sum \operatorname{Ind}(y_H), \quad \text{with } y_H = x_H \cdot \operatorname{Res}_H(y) \in R_{K'}(H).$$

The character of y_H is zero on the p-singular elements of H. If we knew that y_H belonged to $P_{A'}(H)$, it would follow that y belongs to $P_{A'}(G)$. Hence, we are reduced to the case where G *is elementary*.

Now decompose $G = S \times P$ as above. We have

$$R_{K'}(G) = R_{K'}(S) \otimes R_{K'}(P).$$

Since χ is zero off S, we can write χ in the form $f \otimes r_P$, where f is a class function on S, and r_P is the character of the regular representation of P. If ρ is a character of S, then

$$\langle f \otimes r_P, \rho \otimes 1\rangle = \langle f, \rho\rangle \cdot \langle r_P, 1\rangle = \langle f, \rho\rangle.$$

Since the left-hand side is equal to $\langle \chi, \rho \otimes 1\rangle$, it is an integer; thus $\langle f, \rho\rangle \in \mathbf{Z}$ for all ρ, which proves that f is a virtual character of S. Thus we can write y in the form

$$y = y_S \otimes y_P,$$

with $y_S \in R_{K'}(S)$, and y_P the class of the regular representation of P. Since $y_S \in P_{A'}(S)$ and $y_P \in P_{A'}(P)$, we indeed have $y \in P_{A'}(G)$.

(iii) *Sufficiency* (*second part*)

Keep the notation (ii), and suppose in addition that the character χ of y *has values in* K. We must show that y *belongs to* $P_A(G)$. By (ii), we at least know that $y \in P_{A'}(G)$.

Let r be the degree of the extension K'/K. Every $A'[G]$-module defines an $A[G]$-module by restriction, and this module is projective whenever the original module is. Thus we have a homomorphism

$$\pi\colon P_{A'}(G) \to P_A(G).$$

Put $z = \pi(y)$. *Then* $z = r \cdot y$. Indeed, it suffices to verify this equality in $R_{K'}(G)$, and for this it is enough to show that the character χ_z associated with z is equal to $r \cdot \chi$. But we have

$$\chi_z = \mathrm{Tr}_{K'/K}(\chi),$$

and since χ has values in K, we get $\chi_z = r \cdot \chi$.

Thus $y \in P_{A'}(G)$ and $r \cdot y \in P_A(G)$. But the inclusion $P_A(G) \to P_{A'}(G)$ is a *split* injection, cf. 14.6. Since $r \cdot y$ is divisible by r in $P_{A'}(G)$, the same is true in $P_A(G)$, which means that $y \in P_A(G)$, and completes the proof. □

17.6 Proof of theorem 38

We say that a group G is *p-solvable of height h* if it is a successive extension of h groups which are either of order prime to p or of order a power of p. We want to show that, if K is sufficiently large, then every simple $k[G]$-module lifts to an A-free $A[G]$-module.

We proceed by induction on h (the case $h = 0$ being trivial) and, for groups of height h, by induction on the group *order*.

Let I be *a normal subgroup* of G, of order either prime to p or a power of p, such that G/I has height $h - 1$. Let E be a simple (and thus absolutely simple) $k[G]$-module. If I is a p-group, the subspace E^I of all elements of E left invariant by I is $\neq 0$ and therefore equal to E; thus E is a simple $k[G/I]$-module. By induction it can be lifted to an A-free $A[G/I]$ module, and the result follows in this case.

Suppose now that I has *order prime to p*. Decompose E as a direct sum of *isotypic* $k[I]$-modules (i.e., sums of isomorphic simple modules):

$$E = \oplus E_\alpha,$$

where E_α is an isotypic $k[I]$-module of type $\overline{S}_\alpha$.

The group G permutes the E_α; since E is simple it permutes transitively those which are nonzero. Let E_α be one of these, and let G_α be the subgroup of G formed by all $s \in G$ such that $s(E_\alpha) = E_\alpha$. Then E_α is a $k[G_\alpha]$-module, and it is clear that E is the corresponding *induced* module. If $E_\alpha \neq E$, we have $G_\alpha \neq G$, and the induction hypothesis, applied to G_α, shows that E_α can be lifted; consequently the same is true for E.

We can now assume that E *is an isotypic* $k[\mathrm{I}]$-module of type $\bar{\mathrm{S}}$ where $\bar{\mathrm{S}}$ is a simple $k[\mathrm{I}]$-module. Since I has order prime to p, we can lift $\bar{\mathrm{S}}$ in an essentially unique way to an A-free A[I] module, say S, and it is clear that $\mathrm{K} \otimes \mathrm{S}$ is absolutely simple. By cor. 2 to prop. 16 of 6.5, it follows that dim S divides the order of I; in particular, dim S *is prime to p.*

Now let $s \in \mathrm{G}$, and denote by i_s the automorphism $x \mapsto sxs^{-1}$ of I. Since E is isotypic of type $\bar{\mathrm{S}}$, it follows that $\bar{\mathrm{S}}$ (and hence S) is isomorphic to its transform by i_s. This can be expressed as follows:

Let $\rho\colon \mathrm{I} \to \mathrm{Aut}(\mathrm{S})$ be the homomorphism defining the I-module structure of S, and let U_s be the set of $t \in \mathrm{Aut}(\mathrm{S})$ such that

$$t\rho(x)t^{-1} = \rho(sxs^{-1}) \quad \text{for all } x \in \mathrm{I}.$$

Then U_s *is not empty.*

Let G_1 be the group of all pairs (s, t) with $s \in \mathrm{G}$, $t \in \mathrm{U}_s$. The map $(s, t) \mapsto s$ is a surjective homomorphism $\mathrm{G}_1 \to \mathrm{G}$; its kernel is equal to U_1, which is the *multiplicative group* A^* of A. The group G_1 is thus a *central extension* of G by A^*; it acts on S *via* the homomorphism $(s, t) \mapsto t$.

We shall now replace G_1 by a *finite* group. Let $d = \dim \mathrm{S}$. If $s \in \mathrm{G}$, the elements $\det(t)$, $t \in \mathrm{U}_s$, form a coset of A^* modulo A^{*d}. By enlarging K (which is all right, since it does not change $\mathrm{R}_{\mathrm{K}}(\mathrm{G})$), we may assume that these cosets are all trivial, in other words that each U_s contains an element *of determinant* 1. This being done, let C be the subgroup of A^* formed by all $\det(\rho(x))$, $x \in \mathrm{I}$, and let G_2 be the subgroup of G_1 formed by all (s, t) with $t \in \mathrm{U}_s$ and $\det(t) \in \mathrm{C}$. The group G_2 maps *onto* G; the kernel N of $\mathrm{G}_2 \to \mathrm{G}$ is isomorphic to the subgroup of A^* formed by all a with $a^d \in \mathrm{C}$. Since d and Card (C) are prime to p, we conclude that N *is a cyclic group of order prime to p.*

Denote by $\rho_2\colon \mathrm{G}_2 \to \mathrm{Aut}(\mathrm{S})$ the representation $(s, t) \mapsto t$ of G_2. If I is identified with a subgroup of G_2 by means of $x \mapsto (x, \rho(x))$, we see that the restriction of ρ_2 to I is equal to ρ. Thus we have *extended* ρ, not to G itself, but at least to a *central extension* of G (we have a "projective" representation of G in the sense of Schur). Observe that I is normal in G_2, and that $\mathrm{I} \cap \mathrm{N} = \{1\}$.

Return now to the original $k[\mathrm{G}]$-module E. Let $\mathrm{F} = \mathrm{Hom}^{\mathrm{I}}(\bar{\mathrm{S}}, \mathrm{E})$ and let $u\colon \bar{\mathrm{S}} \otimes \mathrm{F} \to \mathrm{E}$ be the homomorphism which associates with $a \otimes b$ ($a \in \bar{\mathrm{S}}$, $b \in \mathrm{F}$ the element $b(a)$ of E.) From the fact that E is isotypic of type $\bar{\mathrm{S}}$ we deduce easily that u is an *isomorphism* of $\bar{\mathrm{S}} \otimes \mathrm{F}$ onto E.

The group G_2 acts on $\bar{\mathrm{S}}$ through the reduction of ρ_2; it also acts on E *via* $\mathrm{G}_2 \to \mathrm{G}$; hence it acts on F. The isomorphism

$$u\colon \bar{\mathrm{S}} \otimes \mathrm{F} \to \mathrm{E}$$

is compatible with this action of G_2. Thus E, viewed as a $k[\mathrm{G}_2]$-module, can be identified with the tensor product of the $k[\mathrm{G}_2]$-modules $\bar{\mathrm{S}}$ and F. In order to lift E, it thus suffices to lift $\bar{\mathrm{S}}$ and F and take the tensor product of these liftings. We will then get an A-free A[G]-module $\tilde{E}$. Since N has order prime to p and acts trivially on the reduction E of $\tilde{E}$, it will follow that N acts trivially on E (cf. 15.5) and that $\tilde{E}$ can be viewed as an A[G]-module—indeed, we will have lifted E.

Hence it remains to show that F can be lifted (the case of $\bar{S}$ being already settled). But F is a *simple* $k[G_2]$-module (since E is) upon which I acts *trivially* by construction. So we may consider it *as a simple* k[H]-*module*, where $H = G_2/I$.

The group H is a central extension of G/I (which is p-solvable of height $\leqslant h - 1$) by the group N, which is cyclic of order prime to p. If $h = 1$, we have H = N, and the lifting of F is immediate (15.5). If $h \geqslant 2$, the group H/N contains a normal subgroup M/N satisfying the following two conditions:

(a) $H/M = (H/N)/(M/N)$ has height $\leqslant h - 2$.
(b) M/N is either a p-group or a group of order prime to p.

If M/N is a p-group, then since N has order prime to p, M can be written as a product $N \times P$ where P is a p-group. The argument given at the beginning of the proof shows that P acts trivially on F, so F can be viewed as a $k[H/P]$-module. But it is clear that the height of H/P is $\leqslant h - 1$, so F can be lifted by induction. There remains the case where M/N has order prime to p. The order of M is then prime to p, and since H/M has height $\leqslant h - 2$, the height of H is $\leqslant h - 1$, and again induction applies. This completes the proof. □

CHAPTER 18

Modular characters

The results we have been discussing are due, for the most part, to R. Brauer. He stated them in a slightly different language, that of *modular characters*, which we shall now describe.

For simplicity, we assume that K *is sufficiently large*.

18.1 The modular character of a representation

Let G_{reg} be the set of p-regular elements of G, and let m' be the l.c.m. of the orders of elements of G_{reg}. By hypothesis, K contains the group μ_K of m'th roots of unity; moreover, since m' is prime to p, reduction mod. $\mathfrak{m}$ is an *isomorphism* of μ_K onto the group μ_k of m'th roots of unity of the residue field k. For $\lambda \in \mu_k$ we let $\tilde{\lambda}$ denote the element of μ_K whose reduction mod. $\mathfrak{m}$ is λ.

Let E be a $k[G]$-module of dimension n, let $s \in G_{reg}$, and let s_E be the endomorphism of E defined by s. Since the order of s is prime to p, s_E is *diagonalizable*, and its eigenvalues $(\lambda_1, \ldots, \lambda_n)$ belong to μ_k. Put

$$\phi_E(s) = \sum_{i=1}^{i=n} \tilde{\lambda}_i .$$

The function $\phi_E : G_{reg} \to A$ thus defined is called the *modular character* (or *Brauer character*) of E. The following properties are immediate:

(i) We have $\phi_E(1) = n = \dim E$.
(ii) ϕ_E is a *class function* on G_{reg}, that is,

$$\phi_E(tst^{-1}) = \phi_E(s) \quad \text{if } s \in G_{reg} \text{ and } t \in G.$$

(iii) If $0 \to E \to E' \to E'' \to 0$ is an exact sequence of $k[G]$-modules, we have

$$\phi_{E'} = \phi_E + \phi_{E''}.$$

(iv) We have

$$\phi_{E_1 \otimes E_2} = \phi_{E_1} \cdot \phi_{E_2}.$$

(v) If $t \in G$ has p'-component $s \in G_{reg}$, the trace of the endomorphism t_E of E is the reduction mod. $\mathfrak{m}$ of $\phi_E(s)$: we have

$$\mathrm{Tr}(t_E) = \overline{\phi_E(s)},$$

where the bar denotes reduction modulo $\mathfrak{m}$. (This can be seen by observing that the eigenvalues of $(t^{-1}s)_E$ are p^αth roots of unity, hence equal to 1 since k has characteristic p. It follows that the eigenvalues of t_E are the same as those of s_E, whence the desired formula.)

(vi) Let F be a K[G]-module with character χ, let E_1 be a lattice of F stable under G, and let $E = E_1/\mathfrak{m}E_1$ be its reduction mod. $\mathfrak{m}$. Then ϕ_E *is the restriction of* χ *to* G_{reg}. (It is enough to see this when G is cyclic of order prime to p. Moreover, th. 32 shows that ϕ_E does not depend on the choice of a stable lattice E_1. This allows a reduction to the case where E_1 is generated by *eigenvectors* of G, in which case the result is clear.)

(vii) If F is a *projective* $k[G]$-module, and if $\tilde{F}$ is a projective A[G]-module whose reduction is F, we shall denote the character of $\tilde{F}$ (i.e., of the K[G]-module $K \otimes \tilde{F}$) by Φ_F. If E is any $k[G]$-module, we know that $E \otimes F$ is a projective $k[G]$-module, and so $\Phi_{E \otimes F}$ makes sense. We have

$$\Phi_{E\otimes F}(s) = \begin{cases} \phi_E(s)\Phi_F(s) & \text{if } s \in G_{reg} \\ 0 & \text{otherwise,} \end{cases}$$

a formula which can be more concisely written as $\Phi_{E \otimes F} = \phi_E \cdot \Phi_F$, even though ϕ_E is not defined off G_{reg}. (We know that $\Phi_{E\otimes F}(s) = 0$ if $s \notin G_{reg}$, cf. th. 36. And by (vi) the restriction of $\Phi_{E \otimes F}$ to G_{reg} is equal to the modular character of $E \otimes F$, which is $\phi_E \cdot \Phi_F$ by (iv).)

(viii) With the same hypothesis as in (vii), we have

$$\langle F, E \rangle_k = \frac{1}{g} \sum_{s \in G_{reg}} \Phi_F(s^{-1})\phi_E(s) = \langle \phi_E, \Phi_F \rangle,$$

where $g = \mathrm{Card}(G)$. (By definition, $\langle F, E\rangle_k$ is the dimension of the largest subspace H^G of $H = \mathrm{Hom}(F, E)$ which is fixed by G. However, H is a projective $k[G]$-module, so if $\tilde{H}$ is the corresponding projective A[G]-module, we see easily that $\dim_k H^G = \mathrm{rank}_A \tilde{H}^G$. If Φ_H is the character of $K \otimes \tilde{H}$, we have

$$\langle F, E \rangle_k = \langle 1, \Phi_H \rangle = \frac{1}{g} \sum_{s \in G} \Phi_H(s).$$

But H is isomorphic to the tensor product of E and the dual of F. By (vii) we have $\Phi_H(s) = \Phi_F(s^{-1})\phi_E(s)$ for $s \in G_{reg}$, and $\Phi_H(s) = 0$ otherwise. The result follows.)

We note the special case where E is the unit representation:

(ix) The subspace F^G formed by the elements invariant under G has dimension

$$\langle \Phi_F, 1 \rangle = \frac{1}{g} \sum_{s \in G_{reg}} \Phi_F(s).$$

Remark. Property (iii) allows us to define the *virtual modular* character ϕ_x of an arbitrary element x of $R_k(G)$. By (vi), if $x = d(y)$ with $y \in R_G(G)$, then ϕ_x is just the restriction to G_{reg} of the virtual character χ_y of y.

It is possible to give analogous definitions for any *linear algebraic group* G over k (assuming here k algebraically closed, for simplicity). The set G_{reg} is then defined as the set of semisimple elements of G. If E is a linear representation of G, and if $s \in G_{reg}$, then $\phi_E(s)$ is defined to be the sum of the *multiplicative representatives* of the eigenvalues of s_E; the modular character ϕ_E thus defined is a class function on G_{reg} with values in A.

18.2 Independence of modular characters

Recall that S_k denotes the collection of isomorphism classes of simple $k[G]$-modules. The various ϕ_E corresponding to elements E of S_k are called the *irreducible modular characters* of the group G.

Theorem 42 . (R. Brauer). *The irreducible modular characters ϕ_E ($E \in S_k$) form a basis of the K-vector space of class functions on G_{reg} with values in K.*

This can be stated in the following equivalent form:

Theorem 42′. *The map $x \mapsto \phi_x$ extends to an isomorphism of $K \otimes R_k(G)$ onto the algebra of class functions on G_{reg} with values in K.*

These theorems immediately give:

Corollary 1. *Let F and F′ be two $k[G]$-modules with the same modular character. Then $[F] = [F']$ in $R_k(G)$; if F and F′ are semisimple, they are isomorphic.*

Corollary 2. *The kernel of the homomorphism d: $R_K(G) \to R_k(G)$ consists of those elements x whose virtual character χ_x is zero on G_{reg}.*

(Since d is surjective, this gives an explicit description of $R_k(G)$ as a quotient of $R_K(G)$.)

Corollary 3. *The number of classes of simple $k[\mathrm{G}]$-modules is equal to the number of p-regular conjugacy classes of* G.

PROOF OF THEOREM 42.

(a) We prove first that $\phi_{\mathrm{E}}(\mathrm{E} \in \mathrm{S}_k)$ are *linearly independent* over K. Indeed, suppose that we had a relation $\sum a_{\mathrm{E}} \phi_{\mathrm{E}} = 0$, with $a_{\mathrm{E}} \in \mathrm{K}$, not all zero. Multiplying the a_{E} by some element of K, we can assume that they all belong to the ring A, and that at least one does not belong to $\mathfrak{m}$. By reduction mod. $\mathfrak{m}$, we then have

$$\sum_{\mathrm{E} \in \mathrm{S}_k} \bar{a}_{\mathrm{E}} \overline{\phi_{\mathrm{E}}(s)} = 0 \quad \text{for all } s \in \mathrm{G}_{\mathrm{reg}},$$

where the bar denotes reduction mod. $\mathfrak{m}$, and one of the $\bar{a}_{\mathrm{E}}$ is not zero. From formula (v) of the preceding section, we get

$$\sum \bar{a}_{\mathrm{E}} \mathrm{Tr}(t_{\mathrm{E}}) = 0 \quad \text{for all } t \in \mathrm{G},$$

thus also for all $t \in k[\mathrm{G}]$. However, since K is sufficiently large, the modules E are absolutely simple, so by the *density theorem* ([8], §4, no. 2), the homomorphism $k[\mathrm{G}] \to \bigoplus_{\mathrm{E} \in \mathrm{S}_k} \mathrm{End}_k(\mathrm{E})$ is surjective. Now let $\mathrm{E} \in \mathrm{S}_k$ such that $\bar{a}_{\mathrm{E}} \neq 0$, let $u \in \mathrm{End}_k(\mathrm{E})$ have trace 1 (a projection on a line, for example), and let t be an element of $k[\mathrm{G}]$ having image u in $\mathrm{End}_k(\mathrm{E})$ and 0 in $\mathrm{End}_k(\mathrm{E}')$ for $\mathrm{E}' \neq \mathrm{E}$. Then we find that $\bar{a}_{\mathrm{E}} \cdot 1 = 0$, a contradiction.

This part of the proof applies just as well to *linear algebraic groups.*

(b) We have to show that the ϕ_{E} *generate* the vector space of class functions on $\mathrm{G}_{\mathrm{reg}}$. Let f be such a function, and extend it to a class function f' on G. We know that f' can be written in the form $\sum \lambda_i \chi_i$ with $\lambda_i \in \mathrm{K}$ and $\chi_i \in \mathrm{R}_{\mathrm{K}}(\mathrm{G})$. Consequently $f = \sum \lambda_i d(\chi_i)$ where $d(\chi_i)$ is the restriction of χ_i to $\mathrm{G}_{\mathrm{reg}}$. Since each $d(\chi_i)$ is a linear combination of the ϕ_{E}, we obtain the desired result. □

EXERCISES

18.1. (In this exercise we do not assume that G is finite or that k has characteristic $\neq 0$.) Let E and E′ be semisimple $k[\mathrm{G}]$-modules. Assume that, for each $s \in \mathrm{G}$, the polynomials $\det(1 + s_{\mathrm{E}}\mathrm{T})$ and $\det(1 + s_{\mathrm{E}'}\mathrm{T})$ are equal. Show that E and E′ are isomorphic. [Reduce to the case where k is algebraically closed and argue as in part (a) of the proof of th. 42.] As a consequence, show that, if E is semisimple and if all the s_{E} are unipotent, then G acts trivially on E (*Kolchin's theorem*).

18.2. Let H be a subgroup of G, let F be a $k[\mathrm{H}]$-module, and let $\mathrm{E} = \mathrm{Ind}_{\mathrm{H}}^{\mathrm{G}} \mathrm{F}$. Show that the modular character ϕ_{E} of E is obtained from ϕ_{F} by the same formula as in the characteristic zero case.

18.3. What is the spectrum of the ring $R_k(G)$?

18.4. Show that the irreducible modular characters form a basis of the A-module of class functions on G_{reg} with values in A. [Use lemma 8 of 10.3 to show that each class function on G_{reg} with values in A extends to a class function on G which belongs to $A \otimes R_K(G)$.]

18.3 Reformulations

We have just seen that $x \mapsto \phi_x$ defines an isomorphism of $K \otimes R_K(G)$ onto the space of class functions on G_{reg}. On the other hand, the map $K \otimes e$: $K \otimes P_k(G) \to K \otimes R_K(G)$ identifies $K \otimes P_k(G)$ with the vector space of class functions on G zero off G_{reg} (this can be checked, for example, by comparing the dimensions of the two spaces). Tensoring with K, the *cde* triangle becomes:

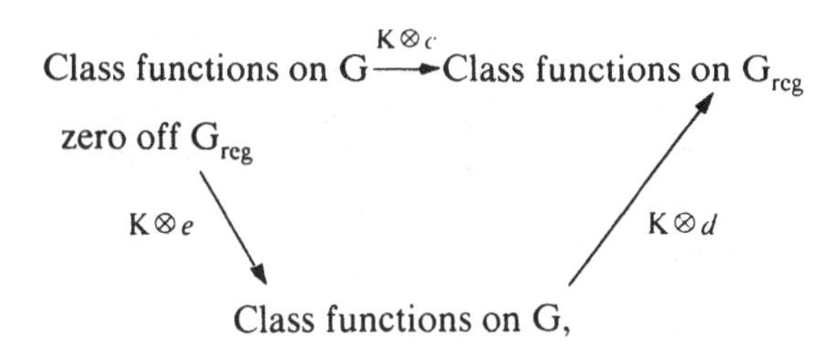

the maps $K \otimes c$, $K \otimes d$, $K \otimes e$ being the obvious ones: restriction, restriction, inclusion. Observe that $K \otimes c$ is an *isomorphism*, in accordance with cor. 1 to th. 35.

The matrices C and D can be interpreted in the following way: if $F \in S_K$, let χ_F denote the character of F; if $E \in S_k$, let ϕ_E denote the modular character of E, and Φ_E the character of the projective envelope of E. Then

$$\chi_F = \sum_{E \in S_k} D_{EF} \phi_E \qquad \text{on } G_{reg}$$

$$\Phi_E = \sum_{F \in S_K} D_{EF} \chi_F \qquad \text{on } G$$

$$\Phi_E = \sum_{E' \in S_k} C_{E'E} \phi_{E'} \qquad \text{on } G_{reg},$$

and we have the orthogonality relations

$$\langle \Phi_E, \phi_{E'} \rangle = \delta_{EE'}, \quad \text{where } \langle \Phi_E, \phi_{E'} \rangle = \frac{1}{g} \sum_{s \in G_{reg}} \Phi_E(s^{-1}) \phi_{E'}(s).$$

We also mention the following version of th. 35:

Theorem 35′. *Let p^n be the largest power of p dividing the order of* G. *If ϕ is a modular character of* G, *and if Φ is defined by the formula*

$$\Phi(s) = p^n\phi(s) \qquad \text{if } s \in \mathrm{G}_{\mathrm{reg}}$$

$$\Phi(s) = 0 \qquad \text{if } s \notin \mathrm{G}_{\mathrm{reg}}$$

then Φ is a virtual character of G.

We leave to the reader the task of making further reformulations of this type.

EXERCISES

18.5. If $s \in \mathrm{G}_{\mathrm{reg}}$, denote by $p^{z(s)}$ the order of a p-Sylow subgroup of the centralizer of s in G.

(a) Let Φ be a class function on G which has values in K. Show that $\Phi \in \mathrm{A} \otimes \mathrm{P}_k(\mathrm{G})$ if and only if Φ is 0 off $\mathrm{G}_{\mathrm{reg}}$ and $\Phi(s) \in p^{z(s)}\mathrm{A}$ for every $s \in \mathrm{G}_{\mathrm{reg}}$ (use ex. 18.4, together with the orthogonality relations $\langle \Phi_{\mathrm{E}}, \phi_{\mathrm{E}'} \rangle = \delta_{\mathrm{EE}'}$).

(b) Use (a) to prove that

$$\mathrm{Coker}(c) \simeq \prod \mathbf{Z}/p^{z(s)}\mathbf{Z} \qquad \text{and} \qquad \det(\mathrm{C}) = p^{\sum z(s)},$$

where s runs through a system of representatives of the p-regular classes of G.

18.6. Assume that G is *p-solvable* (cf. 16.3). If $\mathrm{F} \in \mathrm{S}_{\mathrm{K}}$, let ϕ_{F} denote the restriction of χ_{F} to $\mathrm{G}_{\mathrm{reg}}$. Show that a function ϕ on $\mathrm{G}_{\mathrm{reg}}$ is the modular character of a simple $k[\mathrm{G}]$-module if and only if it satisfies the following two conditions:

(a) There exists $\mathrm{F} \in \mathrm{S}_{\mathrm{K}}$ such that $\phi = \phi_{\mathrm{F}}$.

(b) If $(n_{\mathrm{F}})_{\mathrm{F} \in \mathrm{S}_{\mathrm{K}}}$ is a family of integers $\geqslant 0$ such that $\phi = \sum n_{\mathrm{F}}\phi_{\mathrm{F}}$, then one of the n_{F} is equal to 1 and the others are 0. [Use the Fong–Swan theorem.]

18.4 A section for d

The homomorphism $d\colon \mathrm{R}_{\mathrm{K}}(\mathrm{G}) \to \mathrm{R}_k(\mathrm{G})$ is surjective (th. 33). We shall now describe a *section* for d, i.e., a homomorphism

$$\sigma\colon \mathrm{R}_k(\mathrm{G}) \to \mathrm{R}_{\mathrm{K}}(\mathrm{G})$$

such that $d \circ \sigma = 1$.

For $s \in \mathrm{G}$ let s' denote the p'-component of s. If f is a class function on $\mathrm{G}_{\mathrm{reg}}$, define a class function f' on G by the formula

$$f'(s) = f(s').$$

Theorem 43.

(i) *If f is a modular character of* G, *f' is a virtual character of* G.
(ii) *The map $f \mapsto f'$ defines a homomorphism of* $R_k(G)$ *into* $R_K(G)$ *which is a section for d.*

To prove that f' is a virtual character of G (i.e., belongs to $R_K(G)$), it is enough to prove that, for each elementary subgroup H of G, the restriction of f' to H belongs to $R_K(H)$ (cf. 11.1, th. 21). We are thus reduced to the case where G is *elementary*, and so decomposes as $G = S \times P$ where S has order prime to p and P is a p-group. Moreover, we can assume that f is the modular character of a simple $k[G]$-module E. By the discussion in 15.7, E is even a simple $k[S]$-module, and we can lift it to a simple K[S]-module on which P acts trivially. The character of this module is evidently f', which proves (i).

Assertion (ii) follows from (i) by observing that the restriction of f' to G_{reg} is equal to f. □

EXERCISES

18.7. Let m be the l.c.m. of orders of the elements of G. Write m in the form $p^n m'$ with $(p, m') = 1$ (cf. 18.1.) and choose an integer q such that $q \equiv 0$ (mod. p^n) and $q \equiv 1$ (mod. m').
(a) Show that, if $s \in G$, the p'-component s' of s is equal to s^q.
(b) Let f be a modular character of G, and let ϕ be an element of $R_K(G)$ whose restriction to G_{reg} is f (such an element exists by th. 33). In the notation of th. 43, show that $f' = \Psi^q \phi$, where Ψ^q is the operator defined in ex. 9.3. Deduce from this another proof of the fact that f' belongs to $R_K(G)$ [observe that $R_K(G)$ is stable under Ψ^q].

18.8. Prove th. 43 without assuming K sufficiently large [use the method of the preceding exercise].

18.5 Example: Modular characters of the symmetric group $\mathfrak{S}_4$

The group $\mathfrak{S}_4$ is the group of permutations of $\{a, b, c, d\}$. Recall its character table (cf. 5.8):

	1	(ab)	$(ab)(cd)$	(abc)	$(abcd)$
χ_1	1	1	1	1	1
χ_2	1	-1	1	1	-1
χ_3	2	0	2	-1	0
χ_4	3	1	-1	0	-1
χ_5	3	-1	-1	0	1

We shall determine its irreducible modular characters in characteristic p. We may assume that p divides the order of G, i.e., $p = 2$ or $p = 3$.

(a) *The case* $p = 2$

There are two p-regular classes: that of 1 and that of (abc). By cor. 3 to th. 42, there are two irreducible representations in characteristic 2 (up to isomorphism.) The only representation of degree 1 is the unit representation, with modular character $\phi_1 = 1$. On the other hand, the irreducible representation of degree 2 of $\mathfrak{S}_4$ upon reduction mod. 2 gives a representation ρ_2 whose modular character ϕ_2 takes the value -1 for the element (abc). Consequently, $\bar{\rho_2}$ is not an extension of two representations of degree 1 (otherwise we would have $\phi_2 = 2\phi_1 = 2$), hence is irreducible. The irreducible modular characters of $\mathfrak{S}_4$ are thus ϕ_1 and ϕ_2:

	1	(abc)
ϕ_1	1	1
ϕ_2	2	-1

The *decomposition matrix* D is obtained by expressing the restrictions to G_{reg} of the characters $\chi_1, \ldots, \chi_5$ as a function of ϕ_1 and ϕ_2. We find

$$\chi_1 = \phi_1 \quad \text{on } G_{reg}$$

$$\chi_2 = \phi_1 \quad \text{on } G_{reg}$$

$$\chi_3 = \phi_2 \quad \text{on } G_{reg}$$

$$\chi_4 = \phi_1 + \phi_2 \quad \text{on } G_{reg}$$

$$\chi_5 = \phi_1 + \phi_2 \quad \text{on } G_{reg}$$

hence

$$D = \begin{pmatrix} 1 & 1 & 0 & 1 & 1 \\ 0 & 0 & 1 & 1 & 1 \end{pmatrix}.$$

The characters Φ_1 and Φ_2 of the projective indecomposable modules corresponding to ϕ_1 and ϕ_2 are obtained by means of the transposed matrix of D:

$$\Phi_1 = \chi_1 + \chi_2 + \chi_4 + \chi_5$$

$$\Phi_2 = \chi_3 + \chi_4 + \chi_5 .$$

The corresponding representations have degree 8. The *Cartan matrix* $C = D \cdot {}^tD$ is the matrix $\left(\begin{smallmatrix} 4 & 2 \\ 2 & 3 \end{smallmatrix}\right)$ with determinant 8. It expresses the following decomposition of Φ_1 and Φ_2 on G_{reg}:

$$\Phi_1 = 4\phi_1 + 2\phi_2 \qquad \text{on } G_{reg}$$

$$\Phi_2 = 2\phi_1 + 3\phi_2 \qquad \text{on } G_{reg}.$$

(b) *The case* $p = 3$

There are four p-regular classes: 1, (ab), $(ab)(cd)$, $(abcd)$, hence four irreducible representations in characteristic $p = 3$. On the other hand, the reductions of the characters χ_1, χ_2, χ_4 and χ_5 are irreducible: this is clear for the first two, which have degree 1, and for the two others it follows from the fact that their degree is the largest power of p dividing the group order (cf. 16.4, prop. 46). Since their modular characters are distinct, they are *all* the irreducible modular characters of $\mathfrak{S}_4$. If we denote them by $\phi_1, \phi_2, \phi_3, \phi_4$, we have the table:

	1	(ab)	$(ab)(cd)$	$(abcd)$
ϕ_1	1	1	1	1
ϕ_2	1	-1	1	-1
ϕ_3	3	1	-1	-1
ϕ_4	3	-1	-1	1

Since $\chi_3 = \phi_1 + \phi_2$ on G_{reg} we obtain the following decomposition matrix D and Cartan matrix C:

$$D = \begin{pmatrix} 1 & 0 & 1 & 0 & 0 \\ 0 & 1 & 1 & 0 & 0 \\ 0 & 0 & 0 & 1 & 0 \\ 0 & 0 & 0 & 0 & 1 \end{pmatrix}, \quad C = D \cdot {}^tD = \begin{pmatrix} 2 & 1 & 0 & 0 \\ 1 & 2 & 0 & 0 \\ 0 & 0 & 1 & 0 \\ 0 & 0 & 0 & 1 \end{pmatrix}, \quad \det(C) = 3.$$

The characters $\Phi_1, \ldots, \Phi_4$ of the projective indecomposable modules are:

$$\Phi_1 = \chi_1 + \chi_3$$

$$\Phi_2 = \chi_2 + \chi_3$$

$$\Phi_3 = \chi_4$$

$$\Phi_4 = \chi_5$$

(Note the simple expression of Φ_3 and Φ_4, cf. prop. 46.)

EXERCISES

18.9. Verify the Fong-Swan theorem for $\mathfrak{S}_4$ [check that each ϕ_i is the restriction of some χ_j to G_{reg}].

18.10. Show that the irreducible representations of $\mathfrak{S}_4$ are realizable over the prime field (in any characteristic).

18.11. The group $\mathfrak{S}_4$ has a normal subgroup N of order 4 such that $\mathfrak{S}_4/N$ is isomorphic to $\mathfrak{S}_3$. Show that N acts trivially in each irreducible representation of $\mathfrak{S}_4$ in characteristic 2. Use this to classify such representations.

18.6 Example: Modular characters of the alternating group $\mathfrak{A}_5$

The group $\mathfrak{A}_5$ is the group of even permutations of $\{a, b, c, d, e\}$. It has 60 elements, divided into 5 conjugacy classes:

the identity element 1,
the 15 conjugates of $(ab)(cd)$, which have order 2,
the 20 conjugates of (abc), which have order 3,
the 12 conjugates of $s = (abcde)$, which have order 5,
the 12 conjugates of s^2, which have order 5.

There are 5 irreducible *characters*, given by the following table:

	1	$(ab)(cd)$	(abc)	s	s^2
χ_1	1	1	1	1	1
χ_2	3	-1	0	$z = \dfrac{1+\sqrt{5}}{2}$	z'
χ_3	3	-1	0	$z' = \dfrac{1-\sqrt{5}}{2}$	z
χ_4	4	0	1	-1	-1
χ_5	5	1	-1	0	0

The corresponding representations are:

χ_1: the unit representation

χ_2 and χ_3: two representations of degree 3, realizable over the field $\mathbf{Q}(\sqrt{5})$, and conjugate over $\mathbf{Q}$. They can be obtained by observing that $\{\pm 1\} \times \mathfrak{A}_5$ is a "Coxeter group" with graph $\circ\overset{3}{—}\circ\overset{5}{—}\circ$, and then considering the reflection representation for this group (cf. Bourbaki, *Gr. et Alg. de Lie*, Ch. VI, p. 231, ex. 11).

χ_4: a representation of degree 4, realizable over $\mathbf{Q}$, obtained by removing the unit representation from the permutation representation of $\mathfrak{A}_5$ on $\{a, b, c, d, e\}$, cf. ex. 2.6.

χ_5: a representation of degree 5, realizable over $\mathbf{Q}$, obtained by removing the unit representation from the permutation representation of $\mathfrak{A}_5$ on the set of its 6 subgroups of order 5.

We determine the modular irreducible characters of $\mathfrak{A}_5$ for $p = 2, 3, 5$:

(a) *The case* $p = 2$

There are four p-regular classes, hence 4 modular irreducible characters. Two of these are obvious: the unit character, and the restriction of χ_4 (cf. prop. 46). On the other hand, we have

$$\chi_2 + \chi_3 = 1 + \chi_5 \quad \text{on } G_{\text{reg}},$$

which shows that the reductions of both the irreducible representations of degree 3 are not irreducible (their characters are conjugate over the field $\mathbf{Q}_2$ of 2-adic numbers since $\sqrt{5} \notin \mathbf{Q}_2$). Each must decompose in $R_k(G)$ as a sum of the unit representation and a representation of degree 2, necessarily irreducible. Therefore, the irreducible modular characters $\phi_1, \phi_2, \phi_3, \phi_4$ are given by the table:

	1	(abc)	s	s^2
ϕ_1	1	1	1	1
ϕ_2	2	-1	$z-1$	$z'-1$
ϕ_3	2	-1	$z'-1$	$z-1$
ϕ_4	4	1	-1	-1

We have

$$\chi_1 = \phi_1 \quad \text{on } G_{\text{reg}}$$

$$\chi_2 = \phi_1 + \phi_2 \quad \text{on } G_{\text{reg}}$$

$$\chi_3 = \phi_1 + \phi_3 \quad \text{on } G_{\text{reg}}$$

$$\chi_4 = \phi_4 \quad \text{on } G_{\text{reg}}$$

$$\chi_5 = \phi_1 + \phi_2 + \phi_3 \quad \text{on } G_{\text{reg}}.$$

Whence the matrices D and C:

$$D = \begin{pmatrix} 1 & 1 & 1 & 0 & 1 \\ 0 & 1 & 0 & 0 & 1 \\ 0 & 0 & 1 & 0 & 1 \\ 0 & 0 & 0 & 1 & 0 \end{pmatrix}, \quad C = \begin{pmatrix} 4 & 2 & 2 & 0 \\ 2 & 2 & 1 & 0 \\ 2 & 1 & 2 & 0 \\ 0 & 0 & 0 & 1 \end{pmatrix}, \quad \det(C) = 4.$$

(b) *The case* $p = 3$

One finds 4 irreducible representations in characteristic 3, namely the reductions of the irreducible representations of degree 1, 3, and 4 (two of degree 3). Moreover, we have $\chi_5 = 1 + \chi_4$ on G_{reg}. Hence:

$$D = \begin{pmatrix} 1 & 0 & 0 & 0 & 1 \\ 0 & 1 & 0 & 0 & 0 \\ 0 & 0 & 1 & 0 & 0 \\ 0 & 0 & 0 & 1 & 1 \end{pmatrix}, \quad C = \begin{pmatrix} 2 & 0 & 0 & 1 \\ 0 & 1 & 0 & 0 \\ 0 & 0 & 1 & 0 \\ 1 & 0 & 0 & 2 \end{pmatrix}, \quad \det(C) = 3.$$

(c) *The case* $p = 5$

There are 3 irreducible representations in characteristic 5, the reductions of the irreducible representations of degree 1, 3, and 5 (note that the two representations of degree 3 have isomorphic reductions). Moreover, we have $\chi_4 = \chi_1 + \chi_3$ on G_{reg}. Hence

$$D = \begin{pmatrix} 1 & 0 & 0 & 1 & 0 \\ 0 & 1 & 1 & 1 & 0 \\ 0 & 0 & 0 & 0 & 1 \end{pmatrix}, \quad C = \begin{pmatrix} 2 & 1 & 0 \\ 1 & 3 & 0 \\ 0 & 0 & 1 \end{pmatrix}, \quad \det(C) = 5.$$

EXERCISES

18.12. Check assertions (b) and (c).

18.13. Prove that the irreducible representations of degree 2 of $\mathfrak{A}_5$ in characteristic 2 are realizable over the field $\mathbf{F}_4$ of 4 elements; obtain from this an isomorphism of $\mathfrak{A}_5$ with the group $\mathbf{SL}_2(\mathbf{F}_4)$.

18.14. Show that $\mathfrak{A}_5$ is isomorphic to $\mathbf{SL}_2(\mathbf{F}_5)/\{\pm 1\}$, and use this isomorphism to obtain the list of irreducible representations of $\mathfrak{A}_5$ in characteristic 5.

18.15. Show that χ_5 is monomial, and that χ_2, χ_3, χ_4 are not.

CHAPTER 19

Applications to Artin representations

19.1 Artin and Swan representations

Let E be a field complete with respect to a discrete valuation, let F/E be a finite Galois extension of E, with Galois group G, and assume for simplicity that E and F have the same residue field. If $s \neq 1$ is an element of G and if π is a prime element of F, put

$$i_G(s) = v_F(s(\pi) - \pi),$$

where v_F denotes the valuation of F, normalized so that $v_F(\pi) = 1$.

Put

$$a_G(s) = -i_G(s) \quad \text{if } s \neq 1$$
$$a_G(1) = \sum_{s \neq 1} i_G(s).$$

Clearly v_F is a *class function* on G with integer values. Moreover:

Theorem. *The function a_G is the character of a representation of* G (over a sufficiently large field).

In other words, if χ is any character of G, then the number

$$f(\chi) = \langle a_G, \chi \rangle$$

is a *non-negative integer.*

Using the formal properties of a_G (cf. [25], ch. VI), we see that $f(\chi) \geqslant 0$, and easily reduce the integrality question to the case where G is cyclic (and even, if we like, to the case where G is *cyclic of order a power of the residue characteristic of* E). We can then proceed in several ways:

(i) If χ is a character of degree 1 of G, one shows that $f(\chi)$ coincides with the valuation of the *conductor* of χ in the sense of local class field theory, and this valuation is evidently an integer. This method works, either in the case of a finite residue field (treated initially by Artin) or in the case of an algebraically closed residue field (using a "geometric" analogue of local class field theory); furthermore, the general case follows easily from the case of an algebraically closed residue field.

(ii) The assertion that $f(\chi)$ is an integer is equivalent to certain congruence properties of the "ramification numbers" of the extension F/E. These properties can be proved directly, cf. [25], chap. V, §7, and S. Sen, *Ann. of Math.*, 90, 1969, p. 33–46. □

Now let r_G be the character of the regular representation of G, and put $u_G = r_G - 1$. Let $sw_G = a_G - u_G$. Then

$$sw_G(s) = 1 - i_G(s) \quad \text{if } s \neq 1$$

$$sw_G(1) = \sum_{s \neq 1} (i_G(s) - 1).$$

It is easily checked that, if χ is a character of G, the scalar product $\langle sw_G, \chi \rangle$ is $\geqslant 0$. Using the above theorem, one sees that $\langle sw_G, \chi \rangle$ is a *non-negative* integer for all χ, that is, sw_G is a *character* of G.

The character a_G(resp.sw_G) is called the *Artin* (resp. *Swan*) *character* of the Galois group G; the corresponding representation is called the Artin (resp. Swan) representation of G. An explicit construction of these representations is not known. Nevertheless we can give a simple description of the characters $g \cdot a_G$ and $g \cdot sw_G$, where $g = \text{Card}(G)$:

Let G_i ($i = 0, 1, \ldots$) denote the *ramification groups* of G; thus $s \in G_i$ if and only if $i_G(s) \geqslant i + 1$ or $s = 1$. Put $\text{Card}(G_i) = g_i$. Then one checks that

$$g \cdot a_G = \sum_{i=0}^{\infty} g_i \cdot \text{Ind}_{G_i}^{G}(u_{G_i})$$

and

$$g \cdot sw_G = \sum_{i=1}^{\infty} g_i \cdot \text{Ind}_{G_i}^{G}(u_{G_i})$$

with $u_{G_i} = r_{G_i} - 1$.

In particular we have $sw_G = 0$ if and only if $G_1 = \{1\}$, i.e. the order of G is prime to the residue characteristic of E. (In other words, $sw_G = 0$ if and only if F/E is *tamely ramified*.)

19.2 Rationality of the Artin and Swan representations

Even though a_G and sw_G have values in **Z**, one can give examples where the corresponding representations are not realizable over **Q**, nor even over **R** (cf. [26], §4 and §5). Nevertheless:

Theorem 44. *Let l be a prime number unequal to the residue characteristic of* E.

(i) *The representations of Artin and Swan are realizable over the field $\mathbf{Q}_l$ of l-adic numbers.*
(ii) *There exists a projective $\mathbf{Z}_l[G]$-module Sw_G, unique up to isomorphism, such that $\mathbf{Q}_l \otimes \mathrm{Sw}_G$ has character sw_G.*

It is enough to prove (ii); assertion (i) then follows, since a_G is obtained from sw_G by adding to it u_G, which is realizable over any field.

For this, we apply prop. 44, taking $p = l$, $K = \mathbf{Q}_l$, $n = g = \mathrm{Card}(G)$, and choosing for K′ a sufficiently large finite extension of $\mathbf{Q}_l$. Condition (a) of that proposition is satisfied, cf. 19.1.

To check (b), we use the formula

$$g \cdot sw_G = \sum_{i \geqslant 1} g_i \cdot \mathrm{Ind}_{G_i}^{G}(u_{G_i})$$

given above. By ramification theory, these G_i $(i \geqslant 1)$ have orders prime to l; it follows that every $A'[G_i]$-module is projective (cf. 15.5), where A′ denotes the ring of integers of K′. Hence u_{G_i} is afforded by a projective $A'[G_i]$-module (even by a projective $\mathbf{Z}_l[G_i]$-module if we wish), and the corresponding induced $A'[G]$-module is projective as well. Taking the direct sum of these modules (each repeated g_i times), we obtain a projective $A'[G]$-module with character $g \cdot sw_G$. All the conditions of prop. 44 are thus satisfied, and the theorem follows. □

Remarks

(1) Part (i) of th. 44 is proved in [26] by a somewhat more complicated method, which, however, gives a stronger result: the algebra $\mathbf{Q}_l[G]$ is *quasisplit* (cf. 12.2).

(2) One could get (ii) from (i) combined with the Fong–Swan theorem (th. 38), and with cor. to prop. 45.

(3) There are examples where the Artin and Swan representations are not realizable over $\mathbf{Q}_p$, where p is the residue characteristic of E. However, J.-M. Fontaine has shown (cf. [27]) that these representations are realizable over the field of Witt vectors of e_0, where e_0 denotes the largest subfield of the residue field of E which is algebraic over the prime field.

19.3 An invariant

Let l be a prime number unequal to the residual characteristic of E. Put $k = \mathbf{Z}/l\mathbf{Z}$ and let M be a k[G]-module. We define an invariant $b(\mathrm{M})$ of M by the formula

$$b(\mathrm{M}) = \langle \overline{\mathrm{Sw}_{\mathrm{G}}}, \mathrm{M} \rangle_k = \dim \mathrm{Hom}^{\mathrm{G}}(\overline{\mathrm{Sw}_{\mathrm{G}}}, \mathrm{M}) = \dim \mathrm{Hom}_{\mathbf{Z}_l[\mathrm{G}]}(\mathrm{Sw}_{\mathrm{G}}, \mathrm{M}),$$

where $\overline{\mathrm{Sw}_{\mathrm{G}}} = \mathrm{Sw}_{\mathrm{G}}/l \cdot \mathrm{Sw}_{\mathrm{G}}$ denotes the reduction mod. l of the $\mathbf{Z}_l[\mathrm{G}]$-module Sw_{G} defined by th. 44. The scalar product $\langle \overline{\mathrm{Sw}_{\mathrm{G}}}, \mathrm{M} \rangle_k$ makes sense, since $\overline{\mathrm{Sw}_{\mathrm{G}}}$ is projective, cf. 14.5.

The invariant $b(\mathrm{M})$ has the following properties:

(i) If $0 \rightarrow \mathrm{M}' \rightarrow \mathrm{M} \rightarrow \mathrm{M}'' \rightarrow 0$ is an exact sequence of k[G]-modules, then $b(\mathrm{M}) = b(\mathrm{M}') + b(\mathrm{M}'')$.

(ii) If ϕ_{M} denotes the modular character of M, then

$$b(\mathrm{M}) = \langle sw_{\mathrm{G}}, \phi_{\mathrm{M}} \rangle = \frac{1}{g} \sum_{s \in \mathrm{G}_{\mathrm{reg}}} sw_{\mathrm{G}}(s^{-1})\phi_{\mathrm{M}}(s),$$

cf. 18.1, formula (viii).

(iii)
$$b(\mathrm{M}) = \sum_{i=1}^{\infty} \frac{g_i}{g} \dim_k(\mathrm{M}/\mathrm{M}^{\mathrm{G}_i})$$

where $\mathrm{M}^{\mathrm{G}_i}$ denotes the largest subspace of M fixed by the ith ramification group G_i.

(This follows from the formula $g \cdot sw_{\mathrm{G}} = \sum_{i \geqslant l} g_i \mathrm{Ind}_{\mathrm{G}_i}^{\mathrm{G}}(u_{\mathrm{G}_i})$ by observing that $\langle \mathrm{Ind}_{\mathrm{G}_i}^{\mathrm{G}}(u_{\mathrm{G}_i}), \phi_{\mathrm{M}} \rangle$ is equal to $\dim_k(\mathrm{M}/\mathrm{M}^{\mathrm{G}_i})$ if $i \geqslant 1$.)

(iv) We have $b(\mathrm{M}) = 0$ if and only if G_1 acts trivially on M, i.e., the action of G on M is "tame." [This follows from (iii).]

Thus $b(\mathrm{M})$ measures the "wild ramification" of the module M. This invariant enters into many questions: cohomology of algebraic curves, local factors of zeta functions, conductors of elliptic curves (cf. [28], [29], [30]).

Appendix

Artinian rings

A ring A is said to be *artinian* if it satisfies the following equivalent conditions (cf. Bourbaki, *Alg.* Ch. VIII, §2):

(a) Every decreasing sequence of left ideals of A is stationary.
(b) The left A-module A has finite length.
(c) Every finitely generated left A-module has finite length.

If A is artinian, its radical $\mathfrak{r}$ is nilpotent, and the ring $S = A/\mathfrak{r}$ is semisimple. The ring S can be decomposed as a product $\prod S_i$ of simple rings; each S_i is isomorphic to a matrix algebra $\mathbf{M}_{n_i}(D_i)$ over a (skew) field D_i, and possesses a unique simple module E_i, which is a D_i^0-vector space of dimension n_i. Every semisimple A-module is annihilated by $\mathfrak{r}$ and thus may be viewed as an S-module; if the module is simple, it is isomorphic to one of the E_i.

EXAMPLE. An algebra of finite dimension over a field k is an artinian ring; this applies in particular to the algebra $k[G]$ of a finite group G.

Grothendieck groups

Let A be a ring, and let $\mathfrak{F}$ be a category of left A-modules. The Grothendieck group of $\mathfrak{F}$, denoted $K(\mathfrak{F})$, is the abelian group defined by generators and relations as follows:
Generators. A generator [E] is associated with each $E \in \mathfrak{F}$.
Relations. The relation $[E] = [E'] + [E'']$ is associated with each exact sequence

$$0 \to E \to E' \to E'' \to 0 \quad \text{where } E, E', E'' \in \mathfrak{F}.$$

If H is an abelian group, the homomorphisms $f\colon K(\mathcal{F}) \to H$ correspond bijectively with maps $\phi\colon \mathcal{F} \to H$ which are "additive," i.e., such that $\phi(E) = \phi(E') + \phi(E'')$ for each exact sequence of the above type.

The two most common examples are those where $\mathcal{F}$ is the category of all finitely generated A-modules, or all finitely generated projective A-modules.

Projective modules

Let A be a ring, and P be a left A-module. We say that P is projective if it satisfies the following equivalent conditions (cf. Bourbaki, *Alg.*, Ch. II, §2):

(a) There exists a free A-module of which P is a direct factor.
(b) For every surjective homomorphism $f\colon E \to E'$ of left A-modules, and for every homomorphism $g'\colon P \to E'$, there exists a homomorphism $g\colon P \to E$ such that $g' = f \circ g$.
(c) The functor $E \mapsto \mathrm{Hom}_A(P, E)$ is exact.

In order that a left ideal $\mathfrak{a}$ of A be a direct factor of A as a module, it is necessary and sufficient that there exist $e \in A$ with $e^2 = e$ and $\mathfrak{a} = Ae$; such an ideal is a projective A-module.

Discrete valuations

Let K be a field, and let K^* be the multiplicative group of nonzero elements of K. A discrete valuation of K (cf. [25]) is a surjective homomorphism $v\colon K^* \to Z$ such that

$$v(x + y) \geqslant \mathrm{Inf}(v(x), v(y)) \quad \text{for } x, y \in K^*.$$

Here v is extended to K by setting $v(0) = +\infty$.

The set A of elements $x \in K$ such that $v(x) \geqslant 0$ is a subring of K, called the valuation ring of v (or the ring of *integers* of K). It has a unique maximal ideal, namely the set $\mathfrak{m}$ of all $x \in K$ such that $v(x) \geqslant 1$. The field $k = A/\mathfrak{m}$ is called the residue field of A (or of v).

In order that K be *complete* with respect to the topology defined by the powers of $\mathfrak{m}$, it is necessary and sufficient that the canonical map of A into the projective limit of the $A/\mathfrak{m}^n$ be an isomorphism.

Bibliography: Part III

For modular representations, see Curtis and Reiner [9] and:

[18] R. Brauer. *Über die Darstellung von Gruppen in Galoisschen Feldern. Act. Sci. Ind., 195* (1935).

[19] R. Brauer. Zur Darstellungstheorie der Gruppen endlicher Ordnung. *Math. Zeit., 63* (1956), p. 406–444.

[20] W. Feit. *The Representation Theory of Finite Groups*, North-Holland, 1982.

For Grothendieck groups and their applications to representations of finite groups, see:

[21] R. Swan. Induced representations and projective modules. *Ann. of Math., 71* (1960), p. 552–578.

[22] R. Swan. The Grothendieck group of a finite group. *Topology,* (1963), p. 85–110.

For projective envelopes, see:

[23] M. Demazure and P. Gabriel. *Groupes algébriques*, Tome I, Chapter V, §2, no. 4. Masson and North-Holland, 1970.

[24] I. Giorgiutti. Groupes de Grothendieck. *Ann. Fac. Sci. Univ. Toulouse, 26* (1962), p. 151–207.

For local fields, and the Artin and Swan representations, see:

[25] J.-P. Serre. *Corps Locaux, Act. Sci. Ind.* 1296, Hermann, Paris (1962). (English translation: *Local Fields*, Graduate Texts in Mathematics 67, Springer-Verlag, 1979.)

[26] J.-P. Serre. Sur la rationalité des représentations d'Artin. *Ann. of Math., 72* (1960), p. 406–420.

[27] J.-M. Fontaine. Groupes de ramification et représentations d'Artin. *Ann. Sci. E.N.S., 4* (1971), p. 337–392.

The invariants obtained from the Swan representations are used in:

[28] M. Raynaud. *Caractéristique d' Euler-Poincaré d'un faisceau et cohomologie des variétés abéliennes*. Séminaire Bourbaki, exposé 286, 1964/65, W. A. Benjamin Publishers, New York, 1966.

[29] A. P. Ogg. Elliptic curves and wild ramification. *Amer. J. of Math.*, 89 (1967), p. 1–21.

[30] J.-P. Serre. *Facteurs locaux des fonctions zêta des variétés algébriques* (*définitions et conjectures*). Séminaire Delange-Pisot-Poitou, Paris, 1969/70, exposé 19.

Index of notation

Numbers refer to sections, i.e., "1.1" is Section 1.1.

Index of terminology

Numbers refer to sections, i.e., "1.1" is Section 1.1.

Graduate Texts in Mathematics

(continued from page ii)

66 WATERHOUSE. Introduction to Affine Group Schemes.
67 SERRE. Local Fields.
68 WEIDMANN. Linear Operators in Hilbert Spaces.
69 LANG. Cyclotomic Fields II.
70 MASSEY. Singular Homology Theory.
71 FARKAS/KRA. Riemann Surfaces. 2nd ed.
72 STILLWELL. Classical Topology and Combinatorial Group Theory. 2nd ed.
73 HUNGERFORD. Algebra.
74 DAVENPORT. Multiplicative Number Theory. 3rd ed.
75 HOCHSCHILD. Basic Theory of Algebraic Groups and Lie Algebras.
76 IITAKA. Algebraic Geometry.
77 HECKE. Lectures on the Theory of Algebraic Numbers.
78 BURRIS/SANKAPPANAVAR. A Course in Universal Algebra.
79 WALTERS. An Introduction to Ergodic Theory.
80 ROBINSON. A Course in the Theory of Groups. 2nd ed.
81 FORSTER. Lectures on Riemann Surfaces.
82 BOTT/TU. Differential Forms in Algebraic Topology.
83 WASHINGTON. Introduction to Cyclotomic Fields. 2nd ed.
84 IRELAND/ROSEN. A Classical Introduction to Modern Number Theory. 2nd ed.
85 EDWARDS. Fourier Series. Vol. II. 2nd ed.
86 VAN LINT. Introduction to Coding Theory. 2nd ed.
87 BROWN. Cohomology of Groups.
88 PIERCE. Associative Algebras.
89 LANG. Introduction to Algebraic and Abelian Functions. 2nd ed.
90 BRØNDSTED. An Introduction to Convex Polytopes.
91 BEARDON. On the Geometry of Discrete Groups.
92 DIESTEL. Sequences and Series in Banach Spaces.
93 DUBROVIN/FOMENKO/NOVIKOV. Modern Geometry—Methods and Applications. Part I. 2nd ed.
94 WARNER. Foundations of Differentiable Manifolds and Lie Groups.
95 SHIRYAEV. Probability. 2nd ed.
96 CONWAY. A Course in Functional Analysis. 2nd ed.
97 KOBLITZ. Introduction to Elliptic Curves and Modular Forms. 2nd ed.
98 BRÖCKER/TOM DIECK. Representations of Compact Lie Groups.
99 GROVE/BENSON. Finite Reflection Groups. 2nd ed.
100 BERG/CHRISTENSEN/RESSEL. Harmonic Analysis on Semigroups: Theory of Positive Definite and Related Functions.
101 EDWARDS. Galois Theory.
102 VARADARAJAN. Lie Groups, Lie Algebras and Their Representations.
103 LANG. Complex Analysis. 3rd ed.
104 DUBROVIN/FOMENKO/NOVIKOV. Modern Geometry—Methods and Applications. Part II.
105 LANG. $SL_2(\mathbf{R})$.
106 SILVERMAN. The Arithmetic of Elliptic Curves.
107 OLVER. Applications of Lie Groups to Differential Equations. 2nd ed.
108 RANGE. Holomorphic Functions and Integral Representations in Several Complex Variables.
109 LEHTO. Univalent Functions and Teichmüller Spaces.
110 LANG. Algebraic Number Theory.
111 HUSEMÖLLER. Elliptic Curves.
112 LANG. Elliptic Functions.
113 KARATZAS/SHREVE. Brownian Motion and Stochastic Calculus. 2nd ed.
114 KOBLITZ. A Course in Number Theory and Cryptography. 2nd ed.
115 BERGER/GOSTIAUX. Differential Geometry: Manifolds, Curves, and Surfaces.
116 KELLEY/SRINIVASAN. Measure and Integral. Vol. I.
117 SERRE. Algebraic Groups and Class Fields.
118 PEDERSEN. Analysis Now.
119 ROTMAN. An Introduction to Algebraic Topology.
120 ZIEMER. Weakly Differentiable Functions: Sobolev Spaces and Functions of Bounded Variation.
121 LANG. Cyclotomic Fields I and II. Combined 2nd ed.
122 REMMERT. Theory of Complex Functions. *Readings in Mathematics*
123 EBBINGHAUS/HERMES et al. Numbers. *Readings in Mathematics*
124 DUBROVIN/FOMENKO/NOVIKOV. Modern Geometry—Methods and Applications. Part III.
125 BERENSTEIN/GAY. Complex Variables: An Introduction.
126 BOREL. Linear Algebraic Groups. 2nd ed.
127 MASSEY. A Basic Course in Algebraic Topology.
128 RAUCH. Partial Differential Equations.
129 FULTON/HARRIS. Representation Theory: A First Course. *Readings in Mathematics*
130 DODSON/POSTON. Tensor Geometry.

131 LAM. A First Course in Noncommutative Rings.
132 BEARDON. Iteration of Rational Functions.
133 HARRIS. Algebraic Geometry: A First Course.
134 ROMAN. Coding and Information Theory.
135 ROMAN. Advanced Linear Algebra.
136 ADKINS/WEINTRAUB. Algebra: An Approach via Module Theory.
137 AXLER/BOURDON/RAMEY. Harmonic Function Theory. 2nd ed.
138 COHEN. A Course in Computational Algebraic Number Theory.
139 BREDON. Topology and Geometry.
140 AUBIN. Optima and Equilibria. An Introduction to Nonlinear Analysis.
141 BECKER/WEISPFENNING/KREDEL. Gröbner Bases. A Computational Approach to Commutative Algebra.
142 LANG. Real and Functional Analysis. 3rd ed.
143 DOOB. Measure Theory.
144 DENNIS/FARB. Noncommutative Algebra.
145 VICK. Homology Theory. An Introduction to Algebraic Topology. 2nd ed.
146 BRIDGES. Computability: A Mathematical Sketchbook.
147 ROSENBERG. Algebraic *K*-Theory and Its Applications.
148 ROTMAN. An Introduction to the Theory of Groups. 4th ed.
149 RATCLIFFE. Foundations of Hyperbolic Manifolds.
150 EISENBUD. Commutative Algebra with a View Toward Algebraic Geometry.
151 SILVERMAN. Advanced Topics in the Arithmetic of Elliptic Curves.
152 ZIEGLER. Lectures on Polytopes.
153 FULTON. Algebraic Topology: A First Course.
154 BROWN/PEARCY. An Introduction to Analysis.
155 KASSEL. Quantum Groups.
156 KECHRIS. Classical Descriptive Set Theory.
157 MALLIAVIN. Integration and Probability.
158 ROMAN. Field Theory.
159 CONWAY. Functions of One Complex Variable II.
160 LANG. Differential and Riemannian Manifolds.
161 BORWEIN/ERDÉLYI. Polynomials and Polynomial Inequalities.
162 ALPERIN/BELL. Groups and Representations.
163 DIXON/MORTIMER. Permutation Groups.
164 NATHANSON. Additive Number Theory: The Classical Bases.
165 NATHANSON. Additive Number Theory: Inverse Problems and the Geometry of Sumsets.
166 SHARPE. Differential Geometry: Cartan's Generalization of Klein's Erlangen Program.
167 MORANDI. Field and Galois Theory.
168 EWALD. Combinatorial Convexity and Algebraic Geometry.
169 BHATIA. Matrix Analysis.
170 BREDON. Sheaf Theory. 2nd ed.
171 PETERSEN. Riemannian Geometry.
172 REMMERT. Classical Topics in Complex Function Theory.
173 DIESTEL. Graph Theory. 2nd ed.
174 BRIDGES. Foundations of Real and Abstract Analysis.
175 LICKORISH. An Introduction to Knot Theory.
176 LEE. Riemannian Manifolds.
177 NEWMAN. Analytic Number Theory.
178 CLARKE/LEDYAEV/STERN/WOLENSKI. Nonsmooth Analysis and Control Theory.
179 DOUGLAS. Banach Algebra Techniques in Operator Theory. 2nd ed.
180 SRIVASTAVA. A Course on Borel Sets.
181 KRESS. Numerical Analysis.
182 WALTER. Ordinary Differential Equations.
183 MEGGINSON. An Introduction to Banach Space Theory.
184 BOLLOBAS. Modern Graph Theory.
185 COX/LITTLE/O'SHEA. Using Algebraic Geometry.
186 RAMAKRISHNAN/VALENZA. Fourier Analysis on Number Fields.
187 HARRIS/MORRISON. Moduli of Curves.
188 GOLDBLATT. Lectures on the Hyperreals: An Introduction to Nonstandard Analysis.
189 LAM. Lectures on Modules and Rings.
190 ESMONDE/MURTY. Problems in Algebraic Number Theory.
191 LANG. Fundamentals of Differential Geometry.
192 HIRSCH/LACOMBE. Elements of Functional Analysis.
193 COHEN. Advanced Topics in Computational Number Theory.
194 ENGEL/NAGEL. One-Parameter Semigroups for Linear Evolution Equations.
195 NATHANSON. Elementary Methods in Number Theory.
196 OSBORNE. Basic Homological Algebra.
197 EISENBUD/HARRIS. The Geometry of Schemes.
198 ROBERT. A Course in *p*-adic Analysis.
199 HEDENMALM/KORENBLUM/ZHU. Theory of Bergman Spaces.
200 BAO/CHERN/SHEN. An Introduction to Riemann–Finsler Geometry.

201 HINDRY/SILVERMAN. Diophantine Geometry: An Introduction.
202 LEE. Introduction to Topological Manifolds.
203 SAGAN. The Symmetric Group: Representations, Combinatorial Algorithms, and Symmetric Functions. 2nd ed.
204 ESCOFIER. Galois Theory.
205 FÉLIX/HALPERIN/THOMAS. Rational Homotopy Theory.
206 MURTY. Problems in Analytic Number Theory.
Readings in Mathematics
207 GODSIL/ROYLE. Algebraic Graph Theory.

9 780387 901909